Springer Theses

Recognizing Outstanding Ph.D. Research

Aims and Scope

The series "Springer Theses" brings together a selection of the very best Ph.D. theses from around the world and across the physical sciences. Nominated and endorsed by two recognized specialists, each published volume has been selected for its scientific excellence and the high impact of its contents for the pertinent field of research. For greater accessibility to non-specialists, the published versions include an extended introduction, as well as a foreword by the student's supervisor explaining the special relevance of the work for the field. As a whole, the series will provide a valuable resource both for newcomers to the research fields described, and for other scientists seeking detailed background information on special questions. Finally, it provides an accredited documentation of the valuable contributions made by today's younger generation of scientists.

Theses are accepted into the series by invited nomination only and must fulfill all of the following criteria

- They must be written in good English.
- The topic should fall within the confines of Chemistry, Physics, Earth Sciences, Engineering and related interdisciplinary fields such as Materials, Nanoscience, Chemical Engineering, Complex Systems and Biophysics.
- The work reported in the thesis must represent a significant scientific advance.
- If the thesis includes previously published material, permission to reproduce this must be gained from the respective copyright holder.
- They must have been examined and passed during the 12 months prior to nomination.
- Each thesis should include a foreword by the supervisor outlining the significance of its content.
- The theses should have a clearly defined structure including an introduction accessible to scientists not expert in that particular field.

More information about this series at http://www.springer.com/series/8790

Amit Kumar Singh

Analysis and Design of Power Converter Topologies for Application in Future More Electric Aircraft

Doctoral Thesis accepted by
the National University of Singapore, Singapore

 Springer

Author
Dr. Amit Kumar Singh
Department of Electrical and Computer
  Engineering
National University of Singapore
Singapore

Supervisor
Prof. Sanjib Kumar Panda
Power and Energy Section Electrical
  Machines and Drives Laboratory
  (EMDL), Department of Electrical
  and Computer Engineering
National University of Singapore
Singapore

ISSN 2190-5053   ISSN 2190-5061   (electronic)
Springer Theses
ISBN 978-981-10-8212-2   ISBN 978-981-10-8213-9   (eBook)
https://doi.org/10.1007/978-981-10-8213-9

Library of Congress Control Number: 2018930102

Printed on acid-free paper

This Springer imprint is published by Springer Nature
The registered company is Springer Nature Singapore Pte Ltd.
The registered company address is: 152 Beach Road, #21-01/04 Gateway East, Singapore 189721, Singapore

Supervisor's Foreword

It is a great pleasure to write a foreword to this book on More Electric Aircraft (MEA).

The first aircraft dates back to 2000 years ago in ancient China in the form of a kite. Moving forward, the sky lanterns were invented in the third-century BC in ancient China. There has been tremendous growth in aircraft since then from hot air balloon to fixed-wing aircrafts. In 1903, the Wright brothers invented the first fixed-wing aircraft known as the airplane which led to a new era of transportation. The first widely successful commercial airplane was Boeing 707 which was in service for more than 50 years (1958–2013). Today, the airplanes constitute significant proportions of travel by both people and goods all over the world. By one estimate of the International Air Transport Association (IATA), this year's commercial aircraft will carry nearly four billion passenger which is nearly the double of the passengers 12 years back. This meteoric rise in the number of passengers requires more and more aircrafts.

The increased number of aircrafts poses many new challenges including the environmental pollution and increased fossil fuel consumption. Today, the aircrafts are responsible for 1% of total man-made CO_2 emission which poses a serious threat in the form of global warming. To tackle the above-mentioned challenges, the aircraft industries, similar to car industries, are moving towards electrification of the aircraft by replacing non-electrical parts such as hydraulic, pneumatic and mechanical loads with electrical loads. Such aircrafts are known as MEA. There has been tremendous progress made in MEA in the last decade. Boeing 787, Airbus 380 and Lockheed F35 are the most popular and successful examples of MEA. The adoption of MEA has shown significant improvement in fuel consumption, noise and CO_2 and NOx emission. New research and development suggest that the progress of MEA will pave the path for full electric aircraft in near future where both primary and secondary powers will be electrical power.

But nothing good comes without challenges. The major challenge in the adoption of MEA stems from the increased electrical power consumption. For an MEA such as Boeing 787, the total electrical power consumption is more than 1 MW. Compared to a similar conventional aircraft, it has four times more electrical power.

Increased electrical power requires more conversion and conditioning units for dynamic control and regulation of electrical power. This is where power electronic converters play a pivotal role in enabling aircraft industries to move towards MEA technologies.

Like all other industries, power electronics industry has also made a big leap from vacuum tube based rectifiers to switch mode converters operating in MHz switching frequency range. At a device level, the wide bandgap power devices such as SiC and GaN have become a serious alternative to Si power devices because of their superior switching and thermal characteristics. The development of faster microcontrollers and programmable devices has provided the implementation of complex modulation and control schemes which have been hitherto considered impossible.

The important objective of using power electronic converters for aircraft application is to improve the power density. Weight reduction has significant improvement in the performance of aircraft which is often termed as the snowball effect. However, it must be noted that the improvement in power density should not be at the cost of reliability. The requirements for both power density and reliability provide research opportunities at both power converter topology and device level. Currently, the MEA such as Boeing-787 uses passive kind of power converters. The three main power converter units are Auto Transformer Rectifier Unit (ATRU), Transformer Rectifier Unit (TRU) and Auto Transformer Unit (ATU). The elimination of gearbox-driven units in Boeing-787 results in fixed voltage and variable frequency generator output which is then processed by ATRU, ATU and TRU for different electrical loads in the aircraft. Basically, the ATRU is a non-isolated AC–DC converter whereas the TRU is an isolated AC–DC converter. The ATU is an AC–AC converter which converts variable frequency AC voltage to fixed frequency AC voltage.

This book presents several new power electronic converter topologies suitable for the aircraft system. Power electronic converters with controllable semiconductor switches facilitate the converter operation at high switching frequencies and thus can greatly improve the power density by reducing the passive and magnetic elements. This book covers the shortcomings of existing passive AC–DC converters and proposes novel active AC–DC power converter topologies based on the matrix (3×1) for both non-isolated and isolated AC–DC power conversion. Subsequently, a new Space Vector Modulation (SVM) based switching scheme is proposed and its digital implementation is discussed comprehensively. Finally, a High Voltage (HV) resonant DC–DC converter suitable for smaller aircrafts such as Unmanned Aerial Vehicles (UAVs) is presented. The book extensively covers the theory, analysis, design, simulation and experimental test to validate each of the proposed topologies.

According to me, this book will be beneficial to both industry as well as academic field. Industrial people will get to familiarise with new electrical power conversion techniques, the research communities will get a direction and opportunities to explore the possibilities of further improvement in the proposed power converter topologies. Thus, this book would be useful for all people who are, in anyway, involved in the power electronics domain. Even though the topologies are specific to aircraft application, they can potentially be employed in various similar applications including hybrid and full electric vehicles, energy storage and data centres.

Singapore   Assoc. Prof. Sanjib Kumar Panda
December 2017

Parts of this thesis have been published in the following articles:

Journals

- A. K. Singh, E. Jeyasankar, P. Das and S. K. Panda, "**A Matrix Based Non-isolated Three Phase AC–DC Rectifier With Large Step Down Voltage Gain**," in *IEEE Transactions on Power Electronics*, vol. 32, no. 6, pp. 4796–4811, June 2017. (Published)
- A. K. Singh, E. Jeyasankar, P. Das, S. Panda, "**A Single-Stage Matrix Based Isolated Three Phase AC–DC Converter with Novel Current Commutation**," in *IEEE Transactions on Transportation Electrification*, vol. PP, no.99, pp. 1–1 (Published in Special issue on More Electric Aircraft(MEA))
- A. K. Singh, P. Das and S. K. Panda, "**Analysis and Design of SQR Based High Voltage *LLC* Resonant DC–DC Converter**," in *IEEE Transactions on Power Electronics*, vol. 32, no. 6, pp. 4466–4481, June 2017. (Published)

Conference

- A. K. Singh, E. Jeyasankar, P. Das and S. K. Panda, "**A novel matrix based non-isolated buck-boost converter for more electric aircraft**," *IECON 2016 —42nd Annual Conference of the IEEE Industrial Electronics Society*, Florence, 2016, pp. 1233–1238. (Published)
- A. K. Singh, P. Das and S. K. Panda, "**A high power density three phase AC–DC converter for more electric aircraft (MEA)**," *2015 9th International Conference on Power Electronics and ECCE Asia (ICPE-ECCE Asia)*, Seoul, 2015, pp. 1362–1367. (Published)
- Singh, A. K., Das, P., Panda, S.K., "**A novel matrix based isolated three phase AC–DC converter with reduced switching losses**," *Applied Power Electronics Conference and Exposition (APEC)*, 2015 IEEE, vol., no., pp. 1875,1880, 15–19 March 2015 (Published)
- Singh, A. K., Das, P., Pahlevaninezhad, M., Panda, S. K., "**A novel high output voltage DC–DC *LLC* resonant converter with symmetric voltage quadrupler rectifier for RF communications**," *Telecommunications Energy Conference (INTELEC)*, 2014 IEEE 36th International, vol., no., pp. 1, 6, Sept. 28 2014–Oct. 2 2014 (Published)
- Singh, Panda, S. K., "**Analysis, design and implementation of quadrupler based high voltage full bridge series resonant DC–DC converter**," *Energy Conversion Congress and Exposition (ECCE)*, 2014 IEEE, vol., no., pp. 4059, 4064, 14–18 Sept. 2014 (Published)

Acknowledgements

First and foremost I want to thank my main supervisor Associate Prof. Sanjib Kumar Panda for giving me the opportunity to do my Ph.D. thesis at the Electrical Machines and Drives Laboratory (EMDL). I sincerely appreciate all his contribution of time, numerous ideas and funding to make my Ph.D. experience productive as well as stimulating. As a supervisor, he believes in making his student independent and professionally capable. In addition, I want to thank him for his understanding during hard times. I firmly believe that his guidance has helped me to be a better person both personally and professionally.

I also want to express my sincere appreciation to Dr. Pritam Das for his help and guidance and his interest in my research work. I would also like to thank Dr. Amit Gupta and Dr. Rejeki Simanjorang from Rolls Royce, Singapore for providing valuable feedback in the course of my research.

Many thanks to all my EMDL colleagues for their support, mentorship and friendship. I really enjoyed the wonderful time that I worked here. Although it is not a complete list, I will just mention some of those friends who provided valuable input and constructive criticism to my work directly or indirectly. They are Krishnanand K. R., Satarupa Bal, Dr. Joymala, Kawsar Ali, Jayantika, Sindhu, Dr. Hoang Duc Chinh, Dr. Priyesh Chauhan, Gorla Naga Brahmendra Yadav, Dr. Jeevan Adhikhari, Shiva Muthuraj, Saurabh Bhandari, Prathamesh, Elango, Subash, Sunil Dube, Ravi Kiran, Debjaani, Sandeep Kolluri and so many others.

I would like to thank the administrative staff members of the EMDL, Mr. Y. C. Woo, Mr. M. Chandra and Ms. Nurshaeeda Binte Isa who were always cordial to me and helped me to get things done smoothly. Last but not the least, I want to express my sincere gratitude to my parents, my brothers and sisters for their support and for keeping me motivated.

Declaration

I hereby declare that this thesis is my original work and it has been written by me in its entirety. I have duly acknowledged all the sources of information which have been used in the thesis.

This thesis has also not been submitted for any degree in any university previously.

April 2017

Amit Kumar Singh

Contents

Acronyms

ATRU   Auto Transformer Rectifier Unit
ATU   Auto Transformer Unit
CDR   Current Doubler Rectifier
DPF   Displacement Power Factor
DSP   Digital Signal Processor
ECS   Environment Control System
EPC   Electronic Power Conditioner
EV   Electric Vehicle
FHA   First Harmonic Approximation
FPGA   Field-Programmable Gate Array
HFAC   High-Frequency AC
HVDC   High-Voltage DC
MEA   More Electric Aircraft
MPM   Microwave Power Module
PEC   Power Electronic Converter
PFC   Power Factor Correction
PF   Power Factor
PLL   Phase-Locked Loop
PWM   Pulse Width Modulation
RADAR   Radio Detection And Ranging
RF   Radio Frequency
SPWM   Sinusoidal Pulse Width Modulation
SQR   Symmetrical Quadrupler Rectifier
SSPA   Solid State Power Amplifier
SVM   Space Vector Modulation
THD   Total Harmonic Distortion
TRU   Transformer Rectifier Unit

TWTA	Travelling Wave Tube Amplifier
UAV	Unmanned Aerial Vehicle
ZCS	Zero Current Switching
ZVS	Zero Voltage Switching

List of Figures

List of Tables

Summary

With the aim to improve fuel efficiency and to reduce environmental impacts, aircraft systems are replacing bulky and less efficient non-electrical system with the electrical systems. One of the key benefits of replacing non-electrical system with electrical system is the reduction in total weight of the aircraft. However, more electrical system requires more electrical power to be generated in the aircraft. The increased consumption of electrical power in both civil and military aircrafts has necessitated to use more efficient electrical power conversion technologies. Therefore, power electronic converters play a key role in enabling aircraft industries to move towards more electrical system. In the course of this concept, high-performance power electronic converters especially tailored to meet aircraft specifications are required to improve the overall performance of the aircraft system.

In this work, several power converter topologies suitable for the aircraft system are proposed. Both AC–DC and DC–DC type of the converters are proposed for different electrical loads with an aim to improve the performance of the aircraft system. Comprehensive mathematical analysis and design followed by digital simulation of the power converters are presented. Subsequently, hardware prototypes of each converters are built and experimental tests are carried out to verify the benefits of the proposed solutions.

In the course of this research work, three new matrix-based AC–DC converter topologies suitable for aircraft systems have been proposed. The first topology, a non-isolated matrix based converter with large step-down voltage gain is proposed to replace the conventional three-phase boost or buck rectifiers as front end AC–DC converter. It has been demonstrated that the proposed converter provides improved power conversion efficiency compared to the state-of-the-art three-phase buck rectifier for large step-down voltage gain. The second proposed topology is an isolated converter which provides single-stage conversion without any intermediate DC link capacitor contributing to high power density and high power conversion efficiency. A novel method of adding a series capacitor to reduce duty cycle loss and eliminate voltage spikes has been proposed and subsequently, analysed and implemented both in simulation as well as in hardware. The third proposed topology is a matrix based non-isolated buck–boost converter. It is an extension of non-isolated matrix based buck rectifier and particularly, suited for High Voltage DC (HVDC) bus required in the aircrafts. Being a buck–boost converter, it provides higher power density with lesser complexity in sensing and control. For all the three topologies, comprehensive analysis and design with detailed modes of operation have been presented. Simulation studies followed by experimental test results on laboratory prototypes are demonstrated to validate the suitability of the proposed converter topologies.

To meet the strict input power quality requirements of the aircraft systems, SVM-based switching scheme is proposed for the matrix topology. In the proposed switching scheme, the body diode conduction of switches is avoided resulting in lower switch conduction loss. Moreover, it has been demonstrated that half of the switches in the matrix topology undergo natural Zero Voltage Switching (ZVS) which further improves the power conversion efficiency. To improve the Total Harmonic Distortion (THD) of the input current, SVM-based switching modulation scheme is digitally implemented for higher switching frequency and THD_i below 5% is demonstrated through experimental results. The cost effective and efficient implementation of the switching scheme is carried out using a combination of Digital Signal Processor (DSP) and Field-Programmable Gate Array (FPGA). The combined operation of the DSP and the FPGA provides high-resolution and high-frequency switching signals for the matrix switches.

Further research work is carried with an aim to design and develop a high-voltage DC–DC converter. The proposed converter is developed to power up Microwave Power Module (MPM) based transmitters which are used in smaller aircrafts such as UAVs due to their superior power–weight ratio. An LLC resonant converter with Symmetrical Quadrupler Rectifier (SQR) for 2 kV output at 200 W output power is proposed, analysed, designed and implemented in hardware. The use of SQR reduces the required turns ratio of the high-frequency transformer by multiplying the rectified DC output voltage by four. The LLC converter is operated in discontinuous mode for additional voltage boost from the resonant tank and thus, a new differential equation based method is presented for accurate analysis and design of the proposed converter. The benefits of the proposed analysis method over the usual First Harmonic Approximation (FHA) method have been demonstrated through digital simulation and experimental results.

In the course of this thesis, several new power electronic converter topologies suitable for aircraft system have been proposed, built and tested. Through comprehensive analysis, design and scale down laboratory prototype, the suitability of each power converter has been demonstrated. The proposed topologies are compared with the existing topologies and a comparative evaluation is presented to highlight the benefits and limitations of each proposed topology. The benchmarking of the proposed non-isolated and isolated matrix-based AC–DC converter with respect to passive AC–DC rectifiers and state-of-the-art active AC–DC converters is carried out in terms of power density, power quality and power conversion efficiency.

Chapter 1
Introduction

1.1 Aircraft System

Since 1960, the worldwide air passenger traffic has been growing at an average yearly rate of 9% and it has been estimated that it will continue to grow with a 5–7% rate into the foreseeable future. One obvious reason for such growth is technological advances in aircraft system leading to improved aircraft-efficiency and reduced cost. However, with increased air traffic, the aircraft industries are also facing challenges in terms of CO_2 emission and safety [1]. Today, air transport is responsible for 2% of the total man made CO_2 emission which is estimated to increase further to 3% by 2050. In this regard, the Advisory Council for Aeronautics Research in Europe has set several goals to be achieved by 2020 including 50% reduction of CO_2 emissions; an 80% reduction of NO_X emissions, and a 50% reduction of external noise [2, 3, 3]. Thus, currently, the aircraft industries are driven by three major objectives - 1. improving emissions 2. improving fuel economy and, 3. reducing cost.

The power in an aircraft is generated by gas turbines. A large part of this power is used for propulsion thrust (typically 96% in a passenger aircraft) known as primary or propulsive power. A relatively smaller part (typically 4% in a passenger aircraft) of the total power known as non-propulsive power or secondary power is used to power electrical, mechanical, hydraulic and pneumatic loads. The current demand for more efficient and lighter aircraft has made it necessary to use more electrical system in place of bulky and less efficient non-electrical system. The move towards replacing non-electrical system with more efficient electrical system leads to More Electric Aircraft (MEA). Several aircraft industries such as Boeing and Airbus have developed MEAs. Today, MEAs are applicable to several types of aircrafts including Unmanned Aerial vehicles (UAVs), commercial and military airplanes as they deliver cleaner, quieter and more efficient system [4–8].

Figure 1.1 shows a traditional aircraft power distribution system. In a traditional aircraft the jet engine is connected to gearbox-driven units. The gearbox mechanism is used to drive electrical generator, oil pump, fuel pump and hydraulic pump. The

© Springer Nature Singapore Pte Ltd. 2018
A. K. Singh, *Analysis and Design of Power Converter Topologies for Application in Future More Electric Aircraft*, Springer Theses, https://doi.org/10.1007/978-981-10-8213-9_1

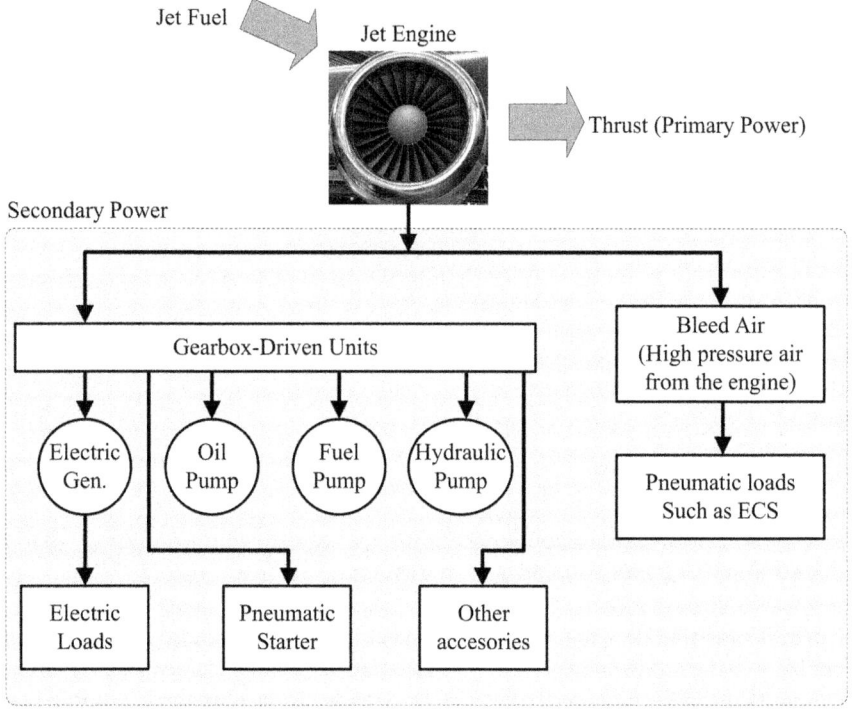

Fig. 1.1 A traditional aircraft system

electrical generator generates AC output power of fixed voltage and fixed frequency. The high pressure air from the engine also known as bleed air is used for pneumatic loads such as Environmental Control System (ECS). In a MEA system, the engine is directly connected to the generators without any gearbox-driven mechanism [9, 10]. Figure 1.2 shows a typical MEA power distribution system. Both AC and DC buses are used in MEAs. For HVDC bus, the three phase AC voltage is rectified to DC voltage using an AC–DC converter. Subsequently, various Power Electronic Converters (PECs) are employed to power different loads of the aircraft system. Apart from the traditional aircraft electric load, in MEA the electric power is required for electric de-icing galley load, ECS, hydraulic and fuel pump motors. Boeing-787, the Dream-liner is an example of MEA which has electric engine start, electric wing ice protection, electric driven hydraulic pumps and ECS [11]. Due to the elimination of pneumatic bleed system from Boeing-787 and use of more electrical system, Boeing-787 have resulted in almost 20% reduction in fuel and CO_2 compared to the conventional Boeing-767 aircraft [12]. Further, the removal of the gearbox driven mechanism in MEA has contributed to significant reduction in weight of the aircraft. Figure 1.3 shows the benefits of Boeing-787 compared to Boeing-767 in terms of reduction in fuel, CO_2 and noise footprint. There is 20% reduction in fuel consumption and CO_2 in Boeing-787 compared to Boeing-767. There is huge reduction of

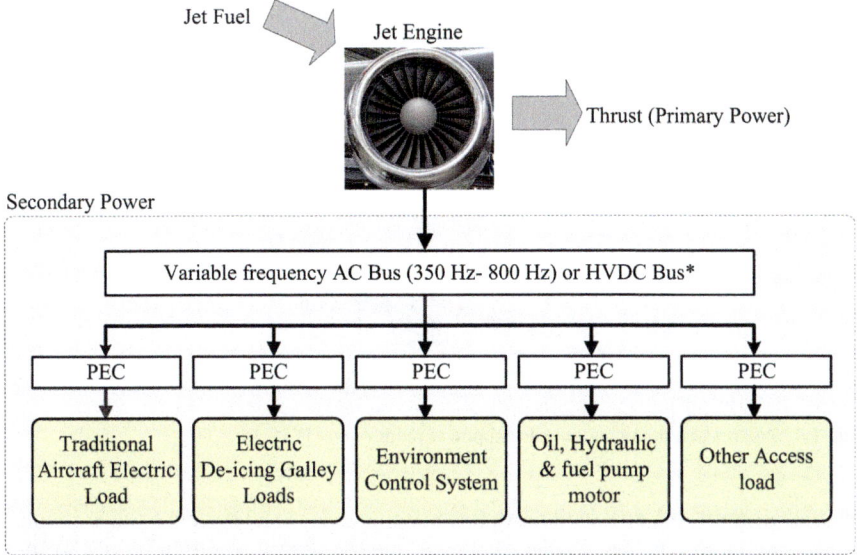

*HVDC bus requires an ac-dc power converter
*PEC – Power Electronic Converter

Fig. 1.2 A typical More Electric Aircraft (MEA) [1]

Fig. 1.3 Benefits of
Boeing-787 relative to
Boeing-767 [12]

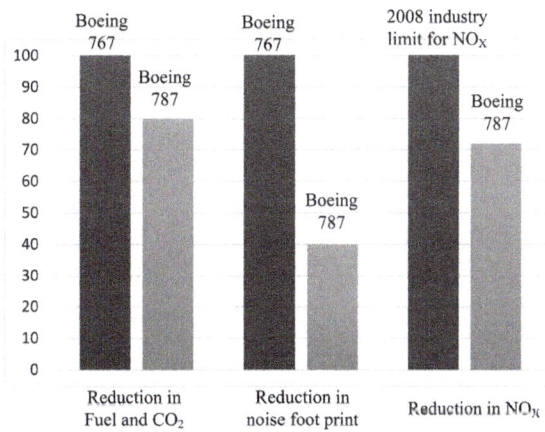

60% in noise foot print of Boeing-787 compared to Boeing-767. Moreover, the NO_X level in Boeing-787 is 28% less than the industry limit for NO_X set in 2008.

As shown in Fig. 1.4, the electrical power consumption in both civil and military aircrafts is growing. For an aircraft of A330/B777 size, the use of more electrical technologies increases the electric power from 200 kW to 1 MW [14]. The military aircrafts are equipped with various electronic subsystem including powerful radars for surveillance and navigation, sensors and weapon systems. Even in com-

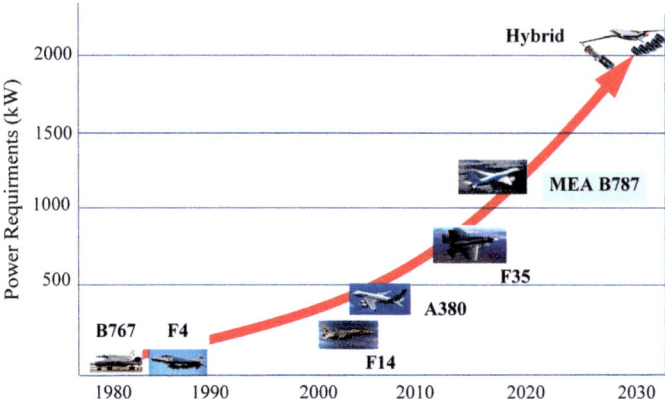

Fig. 1.4 Progression of Aircraft electrical power requirement [13]

mercial aircrafts, the goal is to move towards more electrical system to have simpler, lighter and more efficient system. Weight reduction of the aircraft has huge positive 'snowball effect'.[1] For a short- and mid range aircraft, a reduction of 1 kg weight in equipment will reduce costs by approximately 4500 $ over a 20 years period of operation [15, 16].

One of the enabling technologies for MEA is power electronic converters. Various types of power electronic converters - AC-DC, AC-AC, DC–DC are required for an aircraft system. The two important design objectives of the PEC for aircraft system are *power density* and *reliability*. Moreover, the input power quality for AC–DC converter should also meet DO-160F requirements [1, 17, 18]. Further, the PECs are required to be operated in harsh operating environment. Therefore, it is important that the reliability aspects of the converter is not compromised in process of achieving high power density [18].

1.1.1 Electrical Power System in Commercial Aircrafts

In conventional commercial transport aircrafts, the aircraft generators generate 115-V line-to-neutral AC voltage with a line frequency of 400 Hz. The generators are driven by the main engines. Between the main engine and the generator, a mechanical drive also known as Integrated Drive Generator (IDG) is connected to keep the mechanical speed of the generator constant. The constant mechanical speed results in constant electrical frequency on the aircraft's electric bus.

One of the most distinguishing features of the MEA is the removal IDG. The elimination of IDG results in significant benefits including reduced weight, reduced maintenance cost and increased reliability [18]. However, direct connection of the

---

[1]Describe the multiplication effect in an original weight saving. *Source* wikipedia.

Table 1.1 Key electrical system of Boeing 787 and Airbus 380 [19, 20]

Aircraft	Boeing 787	Airbus 380
No. engines	2	4
No. of generator per engine	2	1
Gen. rating per engine	250 kVA	150 kVA
Gen. output voltage	235 V	115 V
No. of gen. per APU	2	1
Gen. rating per APU	225 kVA	120 kVA
RAT rating	Unavailable	150 kVA
ECS method	Electric compressors	Bleed air
Brake system	Electric	Hydraulic
Actuation System	EHA	Conventional and EHA

main engine with generators requires change in power conversion paradigm due to the variable frequency (350–800 Hz) electric bus. More than ever, the power electronic converters (AC–DC and DC-AC) are required to convert power for many loads including motor drives. As in all applications, power quality is one the most important requirements of the modern aircraft system. The introduction of variable frequency generation presents new challenges for power electronic converters to maintain high power quality and reliability. In contrast to the centralized electrical architecture, MEA favors remote power distribution with new solid-state power controllers and the contactors. The remote power distribution contributes in increased efficiency of the power distribution unit by reducing the distance between generation and consumption. The weight savings and increased efficiency contribute both to improved fuel efficiency and reduced CO_2 emission. All of the above mentioned benefits in MEA have encouraged commercial aircrafts to adopt MEA concepts [4, 5].

The two most well known commercial MEAs are Boeing-787 and Airbus-A380. The key electrical system and their components of the two have been shown in Table 1.1. It is evident that Boeing 787 uses more electrical system compared to Airbus 380. Figure 1.5 shows the electrical power distribution of Boeing-787 [11, 12]. Each of the engines is driving two generators which are generating 230 V AC. The output voltages of generators are regulated using Automatic Voltage Regulator (AVR). However, in the absence of IDG, the output frequency of generators varies from 350 to 800 Hz. An Auxiliary Power Unit (APU) is used to provide axillary power to the aircraft when the aircraft is on ground and main engines are shut down. All the generators of the aircraft are connected to 230 V AC bus as shown in Fig. 1.5.

1.1.1.1 Power Converters in MEA

Aircraft requires various types of power electronic converters for converting electrical power from one voltage level to another as shown in Fig. 1.5. The Auto Transformer

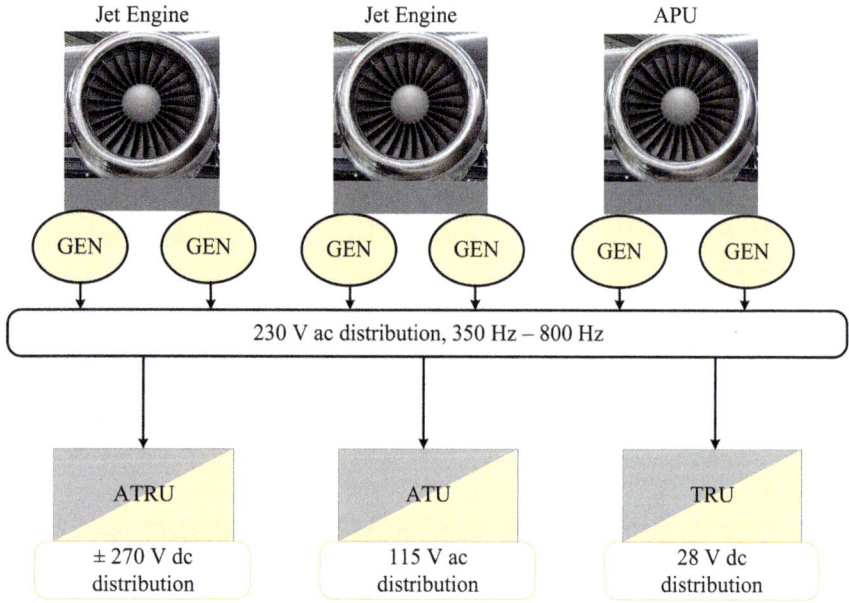

Fig. 1.5 Electrical power distribution in Boeing-787

Rectifier Unit (ATRU) is a non-isolated three phase AC–DC rectifier which converts fixed voltage variable frequency AC generated by the aircraft generators into DC voltage. The Auto Transformer Unit (ATU) is essentially an AC-AC converter which steps down the high voltage AC to low voltage AC. The Transformer Rectifier Unit (TRU) rectifies fixed voltage variable frequency AC voltage into low voltage DC with electrical isolation. This thesis mainly deals with AC–DC rectifiers for the aircraft system [21, 22]. Therefore, the focus is given mainly on ATRU and TRU of the aircraft system. Basically two kinds of AC–DC conversion can be used for aircraft systems - 1. Passive power conversion 2. Active power conversion.

Compared to active power conversion, passive conversion shows benefits in terms of simplicity and reliability. However, the passive power conversion systems such as ATRU and TRU have high weight due to the line frequency (350 Hz) auto-transformer/ transformer. Moreover, the output voltage in passive systems is dependent on the load and mains conditions which is one of the main drawbacks besides weight. The circuit schematic of 12 pulse ATRU is shown in Fig. 1.6. Figure 1.7 shows the dependency of a 12-pulse ATRU output voltage on the output power. The effect of output power on the output voltage is shown for input voltage, $V_{in,rms} = 115$ V and input inductor, $L = 500$ μH [23]. For no load operation, the output voltage is determined by line to line mains voltage whereas for full load the output voltage is minimum at the main frequency, $f = 800$ Hz as shown in Fig. 1.7. The voltage variation must be accommodated in the design of supplied PWM converter or electric machines.

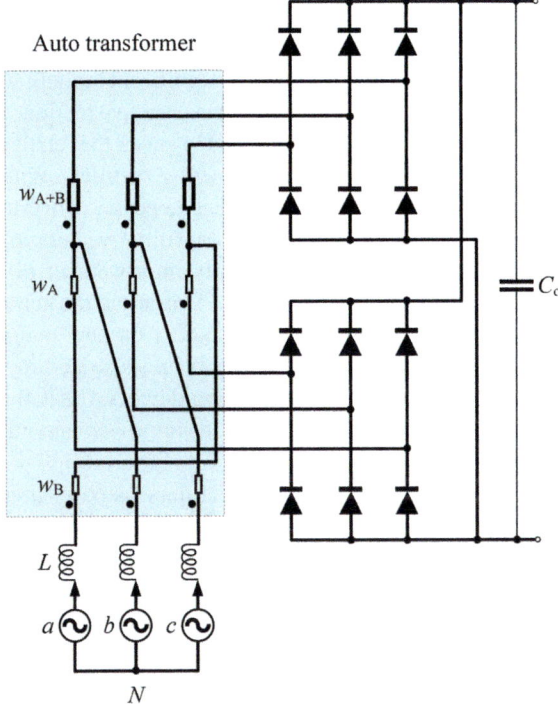

Fig. 1.6 Circuit schematic of a 12 pulse ATRU

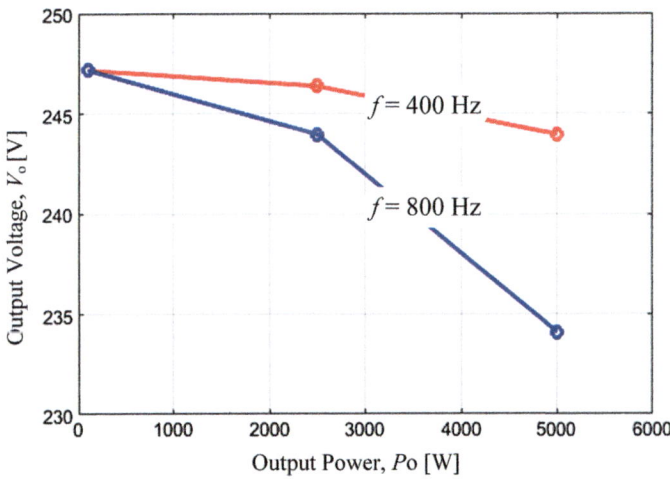

Fig. 1.7 Effect of output power on output voltage for a 12-pulse ATRU at line frequencies, $f = 400$ Hz, 800 Hz. The input rms voltage is 115 V and the value of input inductor, L is 500 μH

The above mentioned disadvantages of the passive system can be overcome by using active power conversion system such as PWM AC–DC rectifier. By operating the active rectifiers at high switching frequency, significant reduction in total system weight can be achieved. The control of the switches in active rectifier allow regulated output voltage with superior input power quality. However, the relative complexity of practical realization and control are still issues in active rectifiers. With digital control system and semiconductor power modules, the active power converters can be good alternative to the passive solutions. With continuous improvement in semiconductor technology, there is great scope in power density improvement of active AC–DC rectifiers by increasing the switching frequency. Contrary to the active rectifiers, the improvement in passive rectifiers is much dependent on new magnetic materials. A comparison of active and passive rectifier has been illustrated in Table 1.2. With current improvements in semiconductor switching devices, the active rectifiers can be of significant advantage for future MEAs with improvements in power density and reliability.

There are various type of active AC–DC rectifiers listed in literature [24–26]. The current passive AC–DC rectifiers ATRU and TRU can be replaced with active non-isolated AC–DC and isolated AC–DC rectifiers, respectively. Moreover, non-isolated AC–DC rectifiers are also required as front end rectifiers for loads such as Electrostatic Hydraulic Actuator (EHA) [23]. Figure 1.8 shows converter-based classification of AC–DC converters. For aircraft system, unidirectional AC–DC converters are required as there is no need of power to flow from load to source [1]. Further, given the input voltage of 115 V or 230 V AC, the buck, boost and buck-boost can provide efficient AC–DC conversion with improved input power quality. As shown in Fig. 1.9, both buck and boost converters have limit over maximum and minimum output voltage. The voltage range which does not fall between the range of buck and boost converter requires buck-boost or boost-buck types of rectifiers. In active AC–DC rectifier, three phase PWM boost converters are widely used nowadays as an alternative to a conventional diode rectifier as they offer several benefits including unity power factor operation, reduced THD at AC mains, constant regulated output

Table 1.2 Comparison between active and passive rectifiers

Active rectifier	Passive rectifier
1. Uses semiconductor switching devices and requires control circuitry.	1. Uses diode based rectifier and does not require control circuitry. Simple and reliable.
2. Does not require any transformer for non-isolated AC–DC power conversion.	2. Requires low frequency auto transformer for non-isolated power conversion
3. Requires high frequency transformer for isolated AC–DC conversion.	3. Requires low frequency transformer for isolated power conversion.
4. Provides controllable power factor.	4. No control over power factor.
5. Switching scheme ensures low THD in input current.	5. Multi-pulse transformers are used to meet the THD requirements
6. Active damping	6. No active damping

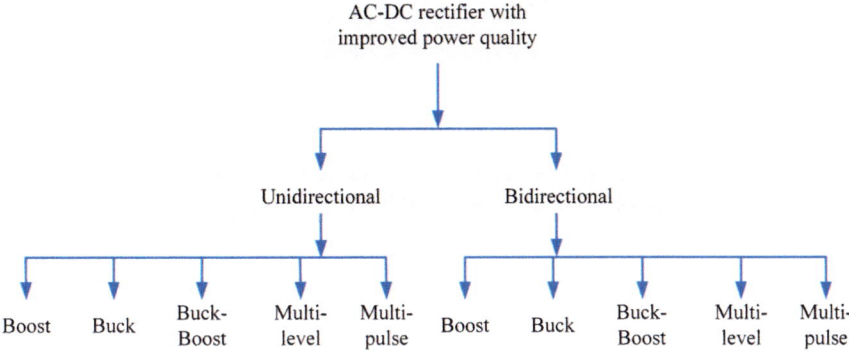

Fig. 1.8 Classification of AC–DC rectifier [24]

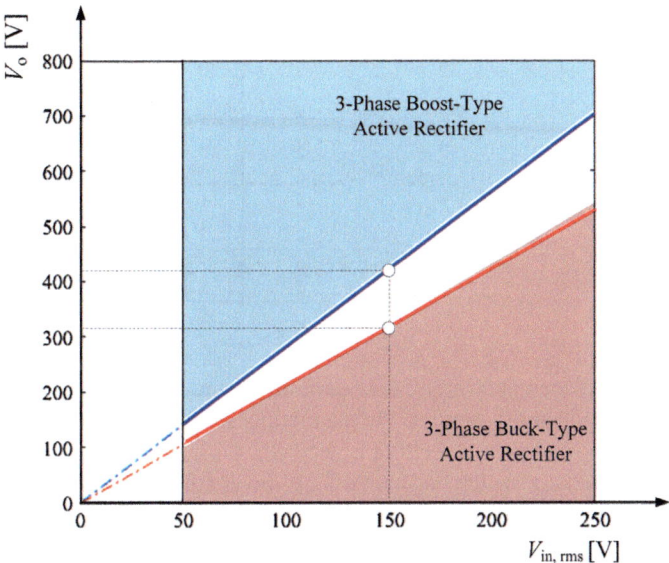

Fig. 1.9 Output voltage range of three phase active rectifiers with boost and buck characteristics. Considering the mains rms voltage, $V_{in,rms}$ to be vary from 0.250 V, the output voltage of the converters are plotted for $m = 1$

DC voltage even under fluctuation of input voltage and output load [27–29]. The two most popular boost rectifiers are, 1. Six switch boost converter [30–33] and 2. Vienna rectifier [34–36]. The six switch boost converter with six switches also known as six-pack power module is widely used in industry because of its simple structure yet high functionality. The circuit schematic of the six switch boost converter is shown in Fig. 1.10a. It offers superior input power quality and is capable of bidirectional power flow. Moreover, the six switch boost converter offers wide range

Fig. 1.10 AC–DC converter topology with boost type characteristic. **a** Circuit topology of six switch boost AC–DC rectifier **b** Circuit topology of original Vienna rectifier

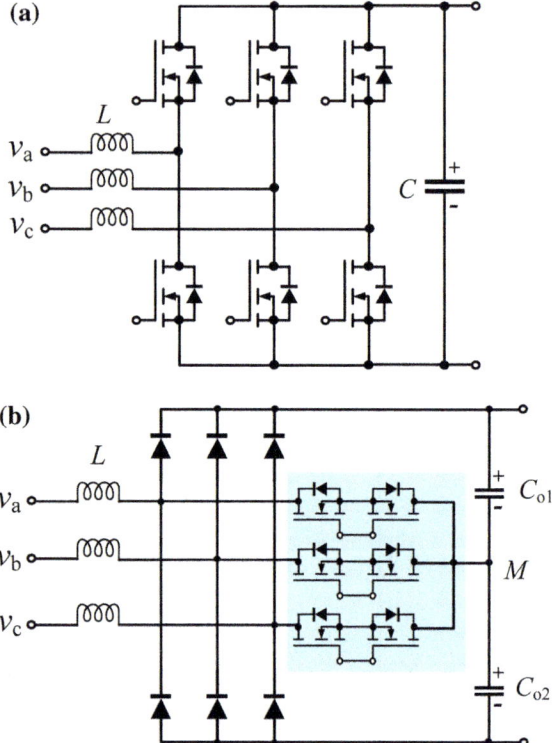

of phase angle displacement which can be exploited for reactive power compensation. Unlike the six switch boost rectifier, Vienna rectifier shows three level characteristic. The circuit topology of Vienna rectifier is shown in Fig. 1.10b. A major advantage of Vienna rectifier over six switch boost rectifier is lower voltage stress across semiconductor switches (half of the voltage stress in six switch boost rectifier). Moreover, it also has lower input mains current ripple resulting in reduced input inductance. The reduction in input inductance translates into higher power density. Also, in Vienna rectifier the short circuit of DC bus because of the faulty control of semi-conductor switches is not an issue which increases the reliability.

For step down output DC voltage, three phase AC–DC rectifier with buck type characteristics are used [37–44]. The buck type of rectifiers have limit over maximum output voltage. The two most popular buck type of AC–DC rectifiers are 1. Active six switch buck type rectifier and 2. Swiss rectifier shown in Fig. 1.11a, b, respectively. In case of active six switch buck type of rectifier, each of the six switches is implemented by series connection of diode and a controllable switch enabling a full voltage-independent capability of the power transfer [40]. By using appropriate modulation scheme such as SPWM and SVM, high input power quality can be guaranteed. Moreover, by using an explicit freewheeling diode across the DC link, the efficiency of the converter can be improved. It should be noted here, that due to use of diode, the

Fig. 1.11 AC–DC converter topology with buck type characteristic. **a** Circuit topology of six switch buck type AC–DC rectifier **b** Circuit topology of Swiss rectifier

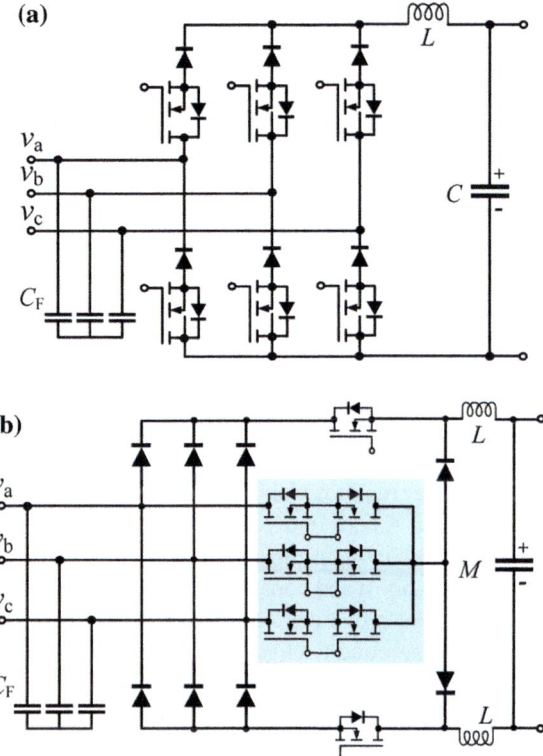

active three phase buck rectifier cannot provide bidirectional power flow. However, by replacing the diode with active semiconductor switch, bidirectional power flow can be ensured. Another main difference of this rectifier from its boost counterpart is that it allows limited phase displacement between input voltage and current and therefore, not very much suitable for reactive power compensation. Swiss rectifier also known as hybrid current injection buck type rectifier provides same output voltage range as of six switch buck rectifier [45, 46]. Though the Swiss rectifier does not provide higher performance as compared to six switch buck rectifier, its DC–DC converter like circuit architecture eases the implementation. In particular, unlike other AC–DC rectifiers, it does not require SVM scheme which typically lead to difficulties in practical implementation. Moreover, like other active rectifiers, it provides sinusoidal input current with controlled output voltage. Besides these two, the three switch buck type of rectifier is also proposed in literature [38]. However, due to higher conduction losses and less uniform distribution of semiconductor losses, it has got limited importance in industries.

For aircraft system, the power converters require buck type of characteristic as the input AC voltage is rectified to relatively lower DC voltage. Moreover, the buck type of rectifiers offer many benefits including smaller line inductors, direct start up without pre-charging and output short circuit protection. Additionally, unlike boost type of rectifiers, buck type of rectifiers do not have any middle point to be stabilized. In terms of sensing and control complexity, the buck type of rectifiers require lesser number of current sensors than boost type of rectifiers. The above mentioned advantages make three phase buck rectifiers a promising topology for aircraft application. In [47], the comparative evaluation of three phase boost rectifier, three phase buck rectifier and three phase Vienna rectifier for 10 kW output power is presented. From the comparison it is observed that the buck rectifier has superior weight which is the main parameter in the aircraft. Regarding reliability and failure tolerance, the buck and Vienna rectifier have approximately the same reliability, whereas the boost is very vulnerable to the transistor shoot-through.

The three phase PWM boost rectifier boosts the input AC voltage and therefore, has a limit over the minimum output voltage. As many aircrafts are moving from 115 V AC bus to 230 V AC bus to reduce the conductor weight [2, 16], the three phase boost converter can not provide the required output voltage. Moreover, the use of boost rectifier requires change in load type which is not a solution aircraft industries would prefer. A buck converter has also limitation over the minimum output voltage due to increase in input current distortion and switch rms current at lower modulation index. Buck-boost and boost-buck topologies provides wider range of output voltage as they provides two degrees of control in forms of modulation index, m and duty cycle, D. The circuit schematic of three phase three switch buck boost converter is shown in Fig. 1.12. Given the choice of buck-boost or boost-buck converter, the buck-boost converter has been found better in terms of power density and control complexity. A comparison between buck-boost and boost-buck converter has been carried out in [50] for 6 kW output power at 25 kHz switching frequency. As shown in Fig. 1.13, the buck -boost converter is half in volume and weight with respect to boost-buck converter.

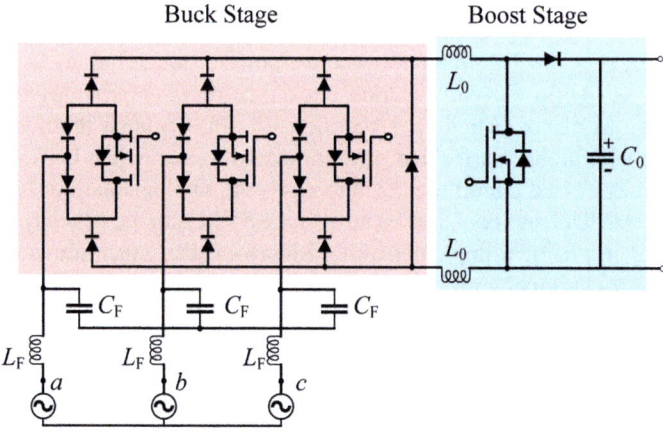

Fig. 1.12 Circuit schematic of the three switch buck-boost converter [48, 49]

Fig. 1.13 Volume and
weight comparison of a
buck-boost and boost-buck
converter [50]

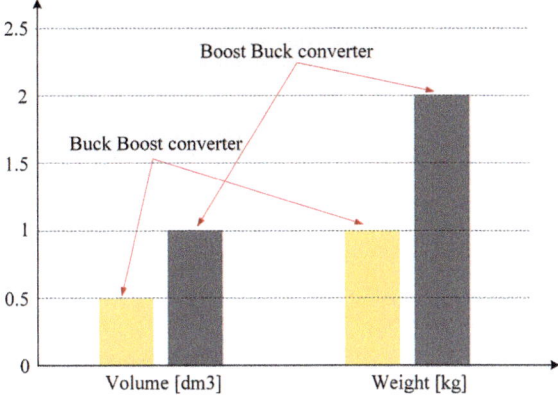

Electrical isolation is preferred for noise sensitive loads and battery charging application. Normally, in aircrafts a line frequency transformer is used to isolate input from output [21, 22]. As the frequency of the AC applied to the transformer can be as low as 350 Hz in aircraft, the size and weight of transformer is bulky. By replacing the line frequency transformer by high frequency transformer, significant weight reduction can be done. Moreover, the use of high frequency transformer facilitates voltage-step down by simply change in the turns ratio of the transformer.

To provide isolated AC–DC power conversion, two stages of conversion, AC–DC followed by isolated DC–DC is normally used in industries for various applications including telecommunication and energy storage [51, 52]. In the first stage, an active Power Factor Correction (PFC) based three phase AC–DC rectifier is used whereas in the second stage, a high frequency isolated DC–DC converted is used. The two stages are linked with intermediated DC link capacitor. Such kind of power converters are also referred to back-to-back power converter. A back-to-back power converter with three phase diode rectifier and boost PFC is shown in Fig. 1.14. Three phase six switch boost rectifier is also used in place of diode rectifier for improved input power quality.

The two stages of conversion can be combined into one by using matrix (3 × 1) topology. The matrix topology directly converts three phase mains frequency AC voltage into single phase high frequency AC voltage which is then processed by high frequency transformer and diode rectifier for output DC voltage [51–53]. The circuit schematic of the matrix based isolated AC–DC rectifier is shown in Fig. 1.15.

Matrix converter has been widely studied for AC-AC conversion and is a promising power converter topology for motor drives, wind energy and solid state transformers [54–56]. Unlike matrix based isolated AC–DC converter, the back-to-back converter requires a mandatory intermediate DC link capacitor. The DC link capacitor used in these converters is often bulky electrolytic capacitor which occupies significant volume, especially for high power applications resulting in poor power density. For power rating of more than 10 kW, the size of DC link capacitor is 30% of the total size of converter. Moreover, the limited life span of the electrolytic capacitor

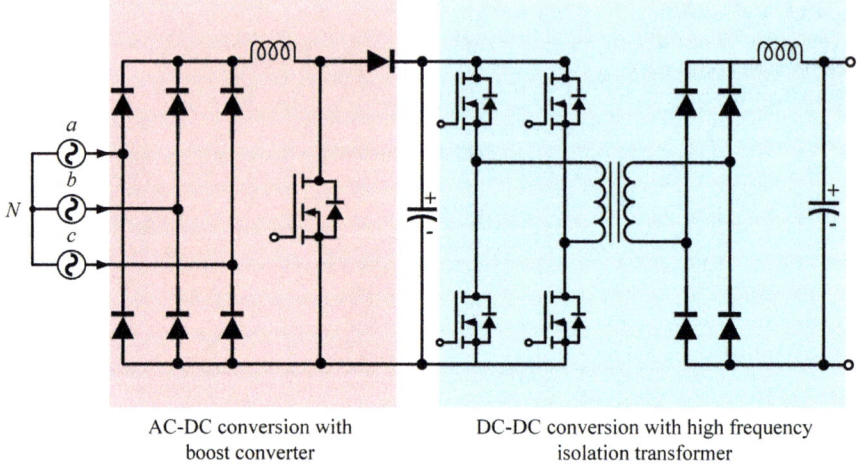

AC-DC conversion with   DC-DC conversion with high frequency
boost converter   isolation transformer

Fig. 1.14 Circuit topology of diode rectifier based isolated AC–DC converter

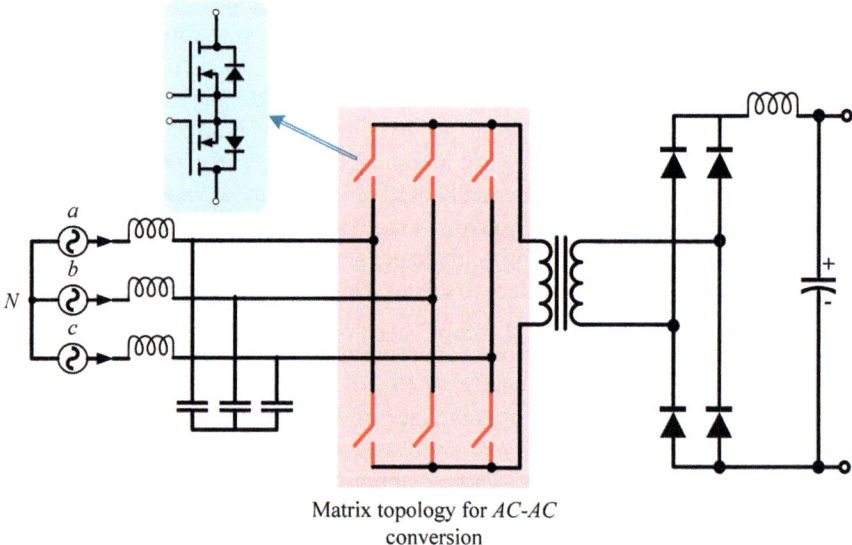

Matrix topology for *AC-AC*
conversion

Fig. 1.15 Circuit topology of matrix based isolated AC–DC converter

presents reliability issues for aerospace applications. In [57] and [58] matrix based converter is compared with the traditional back to back converter. It has been found that the matrix converter provides overall better power density and power conversion efficiency. The elimination of bulky electrolytic capacitor and single stage AC-AC conversion results in improved power density and power conversion efficiency. In [58], the reliability of these two converters are compared. Matrix converter has been found equally reliable than its counterpart back to back converter. Although, the matrix converter has a higher number of semiconductor switches, these switches are subjected to a lower voltage stress, which increases the device reliability. The operation of matrix topology requires four quadrant switch operation which is implemented using connecting two back-to-back MOSFETs as shown in Fig. 1.15. The two back-to-back connected switches allows both positive and negative voltage across the matrix switch in addition to the flow of current in both direction and thus, enables four quadrant operation.

1.1.2   DC–DC Converter for Military Aircraft

The technology planning studies conducted in USA for a military fast jet claims that there are significant aircraft level advantages for more electric technologies including 6.5% reduction in take-off gross weight; 3.2% reduction in life cycle cost; 5.4% increase in 'mean flying hours between failures' and 4.2% reduction in 'maintenance man hours per flying hour' [59]. The MEA approach has been followed in various military aircraft such as F-35 fighter Jet.

One of the main payloads of the military aircraft is powerful RADAR system which is required for various tasks including communication, remote sensing, topographical imaging, surveillance and navigation. Radio Frequency (RF) transmitters are the integral part of a RADAR system. Microwave Power Module (MPM) based RF transmitters are used in such aircrafts because of its superior power to weight ratio compared to its counterparts- Solid State Power Amplifier (SSPA) and Travelling Wave Tube Amplifiers (TWTA). MPMs are hybridization of solid state and vaccum tube technology and used as RF amplifier for medium power application. They have been widely used in Unmanned Aerial Vehicles (UAVs) and satellites owing to their reduced weight and superior RF performance compared to the conventional vaccum tube based transmitter [60–64]. Figure 1.16 shows the comparison of TWTA and MPM based RF transmitter [62, 65]. The operating band of the transmitters

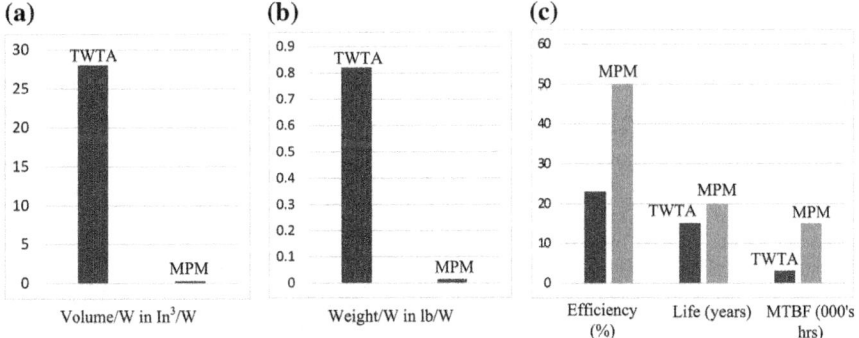

Fig. 1.16 Comparison of TWTA and MPM C-band RF transmitters [62, 65]

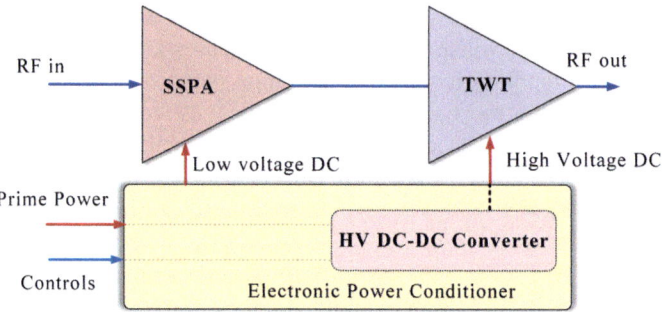

Fig. 1.17 Block diagram of Microwave Power Module (MPM)

is C-Band. As can be seen, the MPM is significantly better in terms of weight/W, volume/W and efficiency which are all very important for military aircraft. Moreover, at the system level the reduced weight and volume of MPM results in reduced complexity and provides better life time and reliability compared to TWTA.

The block diagram of a typical MPM is shown in Fig. 1.17. The RF amplification gain is equally shared between SSPA and TWTA. The power to the SSPA and TWTA is provided by power supply also know as Electronic Power Conditioner (EPC). The EPC consists of several modules including low voltage DC–DC converter, Beam Focusing Electrode (BFE) modulator, high voltage DC–DC converter and communication & control module. Improving the power density of EPC results in reduction of overall transmitter which in turn, reflects in performance improvement of overall aircraft system. One of the main modules of EPC is high voltage DC–DC converter which converts low voltage DC into isolated high voltage DC (in the range of kVs) [63–65]. The two major design objectives of the HV DC–DC converter required for MPMs are 1. reducing the weight and volume, and 2. improving the power conversion efficiency. Resonant converters are suitable to meet these objectives as they can be switched with ZVS /ZCS at very high switching frequency [66–70]. The increased switching frequency allows reduced passive and magnetic elements. Moreover, the use of voltage multiplier instead of rectifier results in reduced turns ratio of step up transformer and thus, in further saving of weight and volume of the converter. Further, the soft switching ability in resonant converters contributes in reduced EMI by reducing $[\frac{di}{dt}]$ and $[\frac{dv}{dt}]$. The block schematic of resonant HV DC–DC converter is shown in Fig. 1.18. First the input DC voltage is converted into high frequency AC using DC-AC converter which can be either full bridge or half bridge. The advantage of half bridge is that it requires only two switches as compared to four switches in full bridge. However, the amplitude of AC output in half bridge is half of the amplitude in full bridge. The high frequency AC is fed to the resonant tank followed by step up high frequency transformer. There are various types of resonant tanks including series, parallel and series-parallel [71, 72]. Subsequently, the stepped up high frequency AC is rectified to high voltage DC using diode rectifiers or voltage multipliers [66, 73–78].

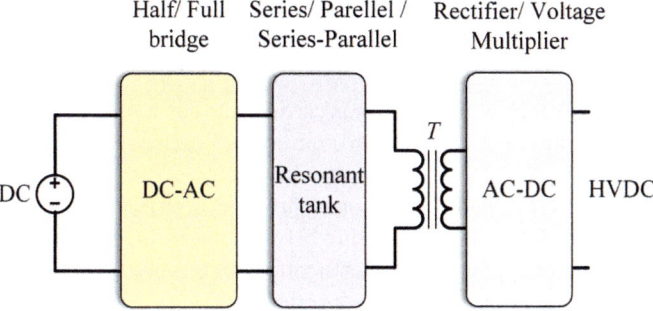

Fig. 1.18 Block diagram of resonant high voltage DC–DC converter

The performance of a MPM based transmitter is very sensitive to the applied high voltage DC [79]. In order to drive TWT, high voltage DC output with very less output voltage ripple is demanded. The output voltage ripple can be reduced by increasing the output filter capacitor. However, the maximum value of the output capacitor is restricted by maximum allowed stored energy in the output side (typically less than 2 J) [61]. In a MPM, the high voltage output is connected to the cathode of the TWT. The phase of the RF output can change almost 100° for 1% change in the cathode voltage [80]. A ripple of 2 V for 2 kV DC output can change the phase by 10°. Therefore, when multiple transmitters are combined, the phase difference results in net RF power loss and this makes it imperative to estimate accurate output voltage ripple during the design of the high voltage DC–DC converter.

1.2 Objective and New Contributions of This Work

The objective of this work is to design and develop high performance power electronic converters suitable for aerospace applications. During the course of this research work, various power electronic converter topologies are proposed, analyzed, simulated and subsequently, validated through hardware experiments. The main focus of this research work are three phase AC–DC (isolated and non-isolated) and high voltage resonant DC–DC converter. The comprehensive analysis and design with comparative evaluation of the proposed PEC topologies with the *state-of-art* power converters are discussed in details.

The main new contributions of this work are as following:

- Proposal of a *novel* matrix (3×1) topology based non-isolated three phase buck rectifier for large step down voltage gain. The proposed non-isolated topology is subsequently, extended for three phase AC–DC conversion with galvanic isolation using high frequency AC transformer. The proposed topology is further configured as three phase non-isolated buck-boost converter for a wide range of output DC voltage.

- Proposal of a new SVM based switching modulation scheme for three phase line frequency AC voltage to single phase high frequency AC conversion. An efficient digital implementation of the switching modulation scheme using micro controller and FPGA based processor is carried out. The proposed switching modulation scheme provides high power conversion efficiency with superior input power quality.
- Small signal modelling and closed loop controller design for the matrix based AC–DC converter. Derivation of relationship between the THD of the input mains current and modulation index. Accurate estimation of the input/output voltage/current ripple for the matrix based AC–DC converter.
- Proposal of an active soft-switched lossless snubber circuit for duty cycle loss minimization and soft commutation of high frequency AC current in the isolated matrix based AC–DC converter. Subsequent validation of the proposed approach using simulation and hardware experiments.
- Proposal and design of SQR based LLC resonant DC–DC converter for high voltage DC output. The proposed converter is operated in discontinuous mode to get additional voltage boost from the resonant tank. Development of an accurate differential equation based analysis and design method for the discontinuous modes of operation of the LLC resonant converter.
- FHA based analysis of SQR based LLC resonant DC–DC converter. Derivation of equivalent load resistance for LLC converter with SQR circuit. The FHA based analysis and design approach is compared with the proposed differential equation based method. The accuracy of the proposed design method over FHA is demonstrated through output voltage ripple estimation.

1.3 Outline of the Thesis

In Chap. 2, a novel matrix-based non-isolated three phase AC–DC converter with a current-doubler rectifier (CDR) circuit is presented. A modified SVM based modulation scheme especially suited for the proposed converter is presented for superior input power quality with reduced power loss. Comprehensive analysis and design of the proposed converter is carried out followed by digital simulation and laboratory based experimental tests. Subsequently, the loss analysis of the proposed converter is carried out and a comparative evaluation of the proposed converter with the traditional three phase buck rectifier is provided to demonstrate the suitability of the proposed converter for large step-down voltage gain. Digital implementation of the proposed modulation scheme is carried out at 40 kHz switching frequency. A laboratory based hardware prototype is developed to validate the theoretical and simulation results.

In Chap. 3, an isolated single-stage three-phase AC–DC converter is proposed for aircraft system. A space vector modulation based switching scheme is proposed and digitally implemented for superior input power quality and improved power conversion efficiency. A current doubler rectifier (CDR) circuit is used at the output side, which essentially reduces the primary to secondary turns ratio. Moreover, an active soft-switched lossless snubber circuit is proposed to overcome the high voltage spikes produced due to the leakage inductance of the high-frequency transformer. A novel approach of adding a series capacitor in the primary winding of the high-frequency transformer for the soft and uninterrupted commutation of the high-frequency AC current is presented. The modes of operation of the proposed converter with the CDR circuit are analyzed in detail. Subsequently, the design equations for the proposed converter are derived. Simulation and experimental results are presented for a laboratory based hardware prototype suitable for aircraft systems.

In Chap. 4, a non-isolated matrix based buck-boost converter applicable for the aircraft systems is presented. Both, the conventional three phase buck and the three phase boost converter have limitations over the maximum and minimum voltage gains, respectively. The proposed buck-boost converter overcomes such limitations and provides output DC voltage with reduced current distortion and improved power conversion efficiency. The benefits and the limitations of the proposed topology over state-of-art three switch buck-boost rectifier are discussed. The topology and the principles of operation of the proposed converter are explained in details. Further, steady state analysis and design of the proposed converter are carried out. A scale down prototype is developed to validate the performance of the proposed buck-boost converter.

In Chap. 5, an isolated high-voltage LLC resonant DC–DC full-bridge converter with SQR circuit is presented. To obtain additional voltage gain from the resonant tank, the converter is operated at switching frequency lower than the series resonant frequency of the LLC tank. A new method based on basic differential equation is proposed for the accurate analysis and design of the converter and subsequently, the key results are compared with a first harmonic approximation based analysis method. A 120 V DC input, 2 kV output, 200 W laboratory prototype has been designed, built, and tested. Simulation and experimental results shown in this chapter demonstrate the validity of the analysis and design of the presented converter.

Chapter 6 summarizes the concluding studies of the work described in this report and some new dimensions are discussed that can be undertaken in future.

1.4 Conclusion

This chapter discusses the electrical distribution system of the traditional aircraft and MEA. The significant benefits of MEA over traditional aircraft in terms of fuel consumption and CO_2 emission are discussed. Subsequently, the power electronics converters required for aircrafts are presented. Various power converter topologies suitable for aircraft are discussed. The benefits and limitation of each topologies are carefully evaluated. The objective and new contribution of the thesis are listed.

References

1. M. Hartmann, *Ultra-Compact and Ultra-Efficient Three-Phase PWM Rectifier Systems for More Electric Aircraft*. PhD thesis, ETH Zurich, 2011
2. K. Rajashekara, Converging technologies for electric/hybrid vehicles and more electric aircraft systems (2010)
3. A. Boglietti, A. Cavagnino, A. Tenconi, S. Vaschetto, P. di Torino, The safety critical electric machines and drives in the more electric aircraft: a survey, in *IECON'09-35th Annual Conference of IEEE Industrial Electronics* (2009), pp. 2587–2594
4. J. Rosero, J. Ortega, E. Aldabas, L. Romeral, Moving towards a more electric aircraft. IEEE Aerosp. Electron. Syst. Mag. **22**, 3–9 (2007)
5. P. Wheeler, S. Bozhko, The more electric aircraft: technology and challenges. IEEE Electrif. Mag. **2**, 6–12 (2014)
6. R. Naayagi, A review of more electric aircraft technology, in *2013 International Conference on Energy Efficient Technologies for Sustainability (ICEETS)* (2013), pp. 750–753
7. R. Jones, The more electric aircraft: the past and the future?, in *Proceedings of the IEE Colloquium on Electrical Machines and Systems for the More Electric Aircraft (Ref. No. 1999/180)* (1999), pp. 1/1–1/4
8. B. Sarlioglu, Advances in ac-dc power conversion topologies for more electric aircraft, in *IEEE Transportation Electrification Conference and Expo (ITEC), 2012* (2012), pp. 1–6
9. J.A. Weimer, The role of electric machines and drives in the more electric aircraft, in *IEEE International Electric Machines and Drives Conference, 2003. IEMDC'03* (2003), vol. 1, pp. 11–15
10. R.E.J. Quigley, More electric aircraft, in *Proceedings Eighth Annual Applied Power Electronics Conference and Exposition* (1993), pp. 906–911
11. G.H. Gaynor, Boeing and the 787 dreamliner (Wiley-IEEE Press, New York, 2015), p. 320
12. K.J. Karimi, Future Aircraft Power Systems- Integration Challenges. https://www.ece.cmu.edu/~electriconf/2008/PDFs/Karimi.pdf
13. D. Koyama, How the More Electric Aircraft is influencing a More Electric Engine and More!. https://sunjet-project.eu/sites/default/files/Rolls-Royce%20-%20Koyama_web.pdf (2015)
14. J. Clare, Examples of More Electric Aircraft Research in the Aerospace Research Centre. http://www.nottingham.ac.uk/aerospace/documents/moreelectricaircarftresearch.pdf
15. J. Brombach, A. Lcken, B. Nya, M. Johannsen, D. Schulz, Comparison of different electrical hvdc-architectures for aircraft application, in *2012 Electrical Systems for Aircraft, Railway and Ship Propulsion (ESARS)* (2012), pp. 1–6
16. J. Brombach, M. Jordan, F. Grumm, D. Schulz, Converter topology analysis for aircraft application, in *International Symposium on Power Electronics Power Electronics, Electrical Drives, Automation and Motion* (2012), pp. 446–451
17. A. Mallik, W. Ding, A. Khaligh, A comprehensive design approach to an emi filter for a 6-kw three-phase boost power factor correction rectifier in avionics vehicular systems, *IEEE Transactions on Vehicular Technology* (2016), vol. PP, no. 99, pp. 1–1
18. B. Sarlioglu, C. Morris, More electric aircraft: review, challenges, and opportunities for commercial transport aircraft. IEEE Trans. Transp. Electrif. **1**, 54–64 (2015)
19. M. Sinnett, 787 No-Bleed Systems: saving Fuel and Enhancing Operational Efficiencies, *Boeing Aero Magazine*, pp. 6–11
20. A.C. (2012) Airbus A350 XWB: the aircraft. http://www.airbus-a350.com/the-aircraft/
21. J. Vieira, J. Oliver, P. Alou, J. Cobos, Power converter topologies for a high performance transformer rectifier unit in aircraft applications, in *2014 11th IEEE/IAS International Conference on Industry Applications (INDUSCON)* (2014), pp. 1–8
22. J. Lee, Aircraft transformer-rectifier units. Stud. Q. J. **42**, 69–71 (1972)
23. G. Gong, M.L. Heldwein, U. Drofenik, J. Minibock, K. Mino, J.W. Kolar, Comparative evaluation of three-phase high-power-factor ac-dc converter concepts for application in future more electric aircraft. IEEE Trans. Ind. Electron. **52**, 727–737 (2005)

24. B. Singh, B. Singh, A. Chandra, K. Al-Haddad, A. Pandey, D. Kothari, A review of three-phase improved power quality ac-dc converters. IEEE Trans. Ind. Electron. **51**, 641–660 (2004)
25. J. Kolar, T. Friedli, The essence of three-phase pfc rectifier systems-part i. IEEE Trans. Power Electron. **28**, 176–198 (2013)
26. T. Friedli, M. Hartmann, J. Kolar, The essence of three-phase pfc rectifier systems-part ii. IEEE Trans. Power Electron. **29**, 543–560 (2014)
27. B.N. Singh, P. Jain, G. Joos, Three-phase ac/dc regulated power supplies: a comparative evaluation of different topologies, in *APEC 2000. Fifteenth Annual IEEE Applied Power Electronics Conference and Exposition (Cat. No.00CH37058)* (2000), vol. 1, pp. 513–518
28. K. Nishimura, K. Atsuumi, K. Tachibana, K. Hirachi, S. Moisseev, M. Nakaoka, Practical performance evaluations on an improved circuit topology of active three-phase pfc power converter, in *APEC 2001. Sixteenth Annual IEEE Applied Power Electronics Conference and Exposition (Cat. No.01CH37181)* (2001), vol. 2, pp. 1308–1314
29. A.W. Green, J.T. Boys, G.F. Gates, 3-phase voltage sourced reversible rectifier, *IEE Proceedings B - Electric Power Applications* (1988), vol. 135, pp. 362–370
30. M. Kumar, L. Huber, M.M. Jovanovi?, Startup procedure for dsp-controlled three-phase six-switch boost pfc rectifier, *IEEE Transactions on Power Electronics* (2015), vol. 30, pp. 4514–4523
31. D. Xu, B. Feng, R. Li, K. Mino, H. Umida, A zero voltage switching svm (zvs ndash;svm) controlled three-phase boost rectifier. IEEE Trans. Power Electron. **22**, 978–986 (2007)
32. S. Hiti, D. Borojevic, R. Ambatipudi, R. Zhang, Y. Jiang, Average current control of three-phase pwm boost rectifier, in *26th Annual IEEE Power Electronics Specialists Conference, 1995. PESC '95 Record* (1995), vol. 1, pp. 131–137
33. Y. Jiang, H. Mao, F.C. Lee, D. Borojevic, Simple high performance three-phase boost rectifiers, in *25th Annual IEEE Power Electronics Specialists Conference, PESC '94 Record* (1994), vol. 2, pp. 1158–1163
34. M. Hartmann, S.D. Round, H. Ertl, J.W. Kolar, Digital current controller for a 1 mhz, 10 kw three-phase vienna rectifier. IEEE Trans. Power Electron. **24**, 2496–2508 (2009)
35. J.W. Kolar, F.C. Zach, A novel three-phase utility interface minimizing line current harmonics of high-power telecommunications rectifier modules. IEEE Trans. Ind. Electron. **44**, 456–467 (1997)
36. P. Karutz, S.D. Round, M.L. Heldwein, J.W. Kolar, Ultra compact three-phase pwm rectifier, in *APEC 07 - Twenty-Second Annual IEEE Applied Power Electronics Conference and Exposition* (2007), pp. 816–822
37. F. Xu, B. Guo, L. Tolbert, F. Wang, B. Blalock, An all-sic three-phase buck rectifier for high-efficiency data center power supplies. IEEE Trans. Ind. Appl. **49**, 2662–2673 (2013)
38. T. Nussbaumer, M. Baumann, J. Kolar, Comprehensive design of a three-phase three-switch buck-type pwm rectifier. IEEE Trans. Power Electron. **22**, 551–562 (2007)
39. B. Guo, F. Wang, R. Burgos, E. Aeloiza, Control of three-phase buck-type rectifier in discontinuous current mode, in *2013 IEEE Energy Conversion Congress and Exposition (ECCE)* (2013), pp. 4864–4871
40. J. Conde-Enriquez, J. Benitez-Read, J. Duran-Gomez, J. Pacheco-Sotelo, Three-phase six-pulse buck rectifier with high quality input waveforms, *IEE Proceedings Electric Power Applications* (1999), vol. 146, pp. 637–645
41. J. Doval-Gandoy, C. Penalver, Dynamic and steady state analysis of a three phase buck rectifier. IEEE Trans. Power Electron. **15**, 953–959 (2000)
42. S.-B. Han, N.-S. Choi, C.-T. Rim, G.-H. Cho, Modeling and analysis of static and dynamic characteristics for buck-type three-phase pwm rectifier by circuit dq transformation. IEEE Trans. Power Electron. **13**, 323–336 (1998)
43. Y. Nishida, T. Kondoh, M. Ishikawa, K. Yasui, Three-phase pwm-current-source type pfc rectifier (theory and practical evaluation of 12kw real product), in *Proceedings of the Power Conversion Conference, 2002. PCC-Osaka 2002* (2002), vol. 3, pp. 1217–1222
44. A. Stupar, T. Friedli, J. Minibock, J.W. Kolar, Towards a 99 % efficient three-phase buck-type pfc rectifier for 400-v dc distribution systems. IEEE Trans. Power Electron. **27**, 1732–1744 (2012)

45. T.B. Soeiro, T. Friedli, J.W. Kolar, Swiss rectifier; a novel three-phase buck-type pfc topology for electric vehicle battery charging, in *2012 Twenty-Seventh Annual IEEE Applied Power Electronics Conference and Exposition (APEC)* (2012), pp. 2617–2624

46. T. Soeiro, T. Friedli, J.W. Kolar, Three-phase high power factor mains interface concepts for electric vehicle battery charging systems, in *2012 Twenty-Seventh Annual IEEE Applied Power Electronics Conference and Exposition (APEC)* (2012), pp. 2603–2610

47. U. Borovi, Analysis and comparison of different active rectifier topologies for avionic specifications, Master's thesis, Universidad Politcnica de Madrid, Spain, 2014

48. M. Baumann, T. Nussbaumer, J.W. Kolar, Comparative evaluation of modulation methods of a three-phase buck + boost pwm rectifier. part i: theoretical analysis. IET Power Electron. **1**, 255–267 (2008)

49. T. Nussbaumer, M. Baumann, J.W. Kolar, Comparative evaluation of modulation methods of a three-phase buck + boost pwm rectifier. part ii: experimental verification. IET Power Electronics **1**, 268–274 (2008)

50. J.W.Kolar, T. Nussbaumer, K. Mino, Design and comparative evaluation of three-phase buck+boost and boost+buck unity power factor pwm rectifier systems for supplying variable dc voltage link converters, in *25th International Conference on Power Electronics (PCIM)* (2004)

51. S. Ratanapanachote, H.J. Cha, P. Enjeti, A digitally controlled switch mode power supply based on matrix converter, in *2004 IEEE 35th Annual Power Electronics Specialists Conference, 2004. PESC 04* (2004), vol. 3, pp. 2237–2243

52. J. Sandoval, S. Essakiappan, P. Enjeti, A bidirectional series resonant matrix converter topology for electric vehicle dc fast charging, in *2015 IEEE Applied Power Electronics Conference and Exposition (APEC)* (2015), pp. 3109–3116

53. G.T. Chiang, K. Orikawa, Y. Ohnuma, J.I. Itoh, Improvement of output voltage with svm in three-phase ac to dc isolated matrix converter, in *IECON 2013 - 39th Annual Conference of the IEEE Industrial Electronics Society* (2013), pp. 4862–4867

54. P. Wheeler, J. Clare, L. Empringham, M. Apap, M. Bland, Matrix converters. Power Eng. J. **16**, 273–282 (2002)

55. P. Wheeler, J. Clare, L. Empringham, M. Apap, K. Bradley, C. Whitley, G. Towers, A matrix converter based permanent magnet motor drive for an aircraft actuation system, in *IEEE International Electric Machines and Drives Conference, 2003. IEMDC'03* (2003), vol. 2, pp. 1295–1300

56. K. Basu, N. Mohan, A single-stage power electronic transformer for a three-phase pwm ac/ac drive with source-based commutation of leakage energy and common-mode voltage suppression. IEEE Trans. Ind. Electron. **61**, 5881–5893 (2014)

57. T. Friedli, J. Kolar, J. Rodriguez, P. Wheeler, Comparative evaluation of three-phase ac-ac matrix converter and voltage dc-link back-to-back converter systems. IEEE Trans. Ind. Electron. **59**, 4487–4510 (2012)

58. M. Aten, G. Towers, C. Whitley, P. Wheeler, J. Clare, K. Bradley, Reliability comparison of matrix and other converter topologies. IEEE Trans. Aerosp. Electron. Syst. **42**, 867–875 (2006)

59. W. Pearson, The more electric/all electric aircraft-a military fast jet perspective, in *Proceedings of the IEE Colloquium on All Electric Aircraft (Digest No. 1998/260)* (1998), pp. 5/1–5/7

60. C. Wan, C. Marotta, A. Zubyk, G. Tucker, C. Meadows, R. True, T. Schoemehl, R. Duggal, M. Kirshner, R. Kowalczyk et al., A 100 watt w-band mpm, in *2013 IEEE 14th International Vacuum Electronics Conference (IVEC)* (IEEE, 2013), pp. 1–1

61. R. Duggal, A. Donald, T. Schoemehl, Technological evolution of the microwave power module (mpm), in *IEEE International Vacuum Electronics Conference, 2009. IVEC '09* (2009), pp. 353–354

62. C.R. Smith, C.M. Armstrong, J. Duthie, The microwave power module: a versatile rf building block for high-power transmitters, in *Proceedings of the IEEE* (1999), vol. 87, pp. 717–737

63. C.D. Prasad, G. Baranidharan, K.B. Venkataraman, U.K. Revankar, Integration and evaluation of a mpm based transmitter on fighter aircraft, in *2011 IEEE International Vacuum Electronics Conference (IVEC)* (2011), pp. 435–436

64. P. Sidharthan, K. Mirjith, A.J. Zabiulla, S. Kamath, Development of a fast switching modulator for an mpm, in *2011 IEEE International Vacuum Electronics Conference (IVEC)* (2011), pp. 433–434

65. T. Ninnis, Microwave Power Modules (MPMs) Miniature Microwave Amplifiers for Radars. http://bbs.hwrf.com.cn/downebd/81825d1350949095-13-mpm-radar.pdf (2005)

66. A. Santoja, A. Barrado, C. Fernandez, M. Sanz, C. Raga, A. Lazaro, High voltage gain dc-dc converter for micro and nanosatellite electric thrusters, in *2013 Twenty-Eighth Annual IEEE Applied Power Electronics Conference and Exposition (APEC)* (IEEE, 2013), pp. 2057–2063

67. T.B. Soeiro, J. Muhlethaler, J. Linner, P. Ranstad, J.W. Kolar, automated design of a high-power high-frequency lcc resonant converter for electrostatic precipitators. IEEE Trans. Ind. Electron. **60**(11), 4805–4819 (2013)

68. R. Casanueva, C. Brañas, F.J. Azcondo, F.J. Diaz, Teaching resonant converters: properties and applications for variable loads. IEEE Trans. Ind. Electron. **57**(10), 3355–3363 (2010)

69. J. Liu, L. Sheng, J. Shi, Z. Zhang, X. He, Design of high voltage, high power and high frequency transformer in lcc resonant converter, in *Twenty-Fourth Annual IEEE Applied Power Electronics Conference and Exposition, 2009. APEC 2009* (IEEE, 2009), pp. 1034–1038

70. S. Gavin, M. Carpita, P. Ecoeur, H.-P. Biner, M. Paolone, E.T. Louokdom, A digitally controlled 125 kvdc, 30kw power supply with an lcc resonant converter working at variable dc-link voltage: full scale prototype test results (2014)

71. J.M. Cyr, K. Al-Haddad, L.A. Dessaint, M. Saad, V. Rajagopalan, Comparison of resonant converter topologies. Can. J. Electr. Comput. Eng. **20**, 193–201 (1995)

72. D.M. Divan, Design considerations for very high frequency resonant mode dc/dc converters, *IEEE Transactions on Power Electronics* (1987), vol. PE-2, pp. 45–54

73. S. Iqbal, A three-phase symmetrical multistage voltage multiplier. IEEE Power Electron. Lett. **3**(1), 30–33 (2005)

74. S. Iqbal, A hybrid symmetrical voltage multiplier. IEEE Trans. Power Electron. **29**(1), 6–12 (2014)

75. Y. Zhao, X. Xiang, W. Li, X. He, C. Xia, Advanced symmetrical voltage quadrupler rectifiers for high step-up and high output-voltage converters. IEEE Trans. Power Electron. **28**(4), 1622–1631 (2013)

76. A. Lamantia, P.G. Maranesi, L. Radrizzani, Small-signal model of the cockcroft-walton voltage multiplier. IEEE Trans. Power Electron. **9**(1), 18–25 (1994)

77. N. Vishwanathan, V. Ramanarayanan, Input voltage modulated high voltage dc power supply topology for pulsed load applications, in *IECON 02 [IEEE 2002 28th Annual Conference of the Industrial Electronics Society]* (2002), vol. 1, pp. 389–394

78. B.S. Nathan, V. Ramanarayanan, Analysis, simulation and design of series resonant converter for high voltage applications, in *Proceedings of IEEE International Conference on Industrial Technology 2000* (IEEE, 2000), vol. 1, pp. 688–693

79. I. Barbi, R. Gules, Isolated dc-dc converters with high-output voltage for twta telecommunication satellite applications. IEEE Trans. Power Electron. **18**, 975–984 (2003)

80. T. Mec, TWT Performance Fundamentals. http://www.teledyne-mec.com/products/technical_description.aspx

Chapter 2
A Matrix Based Non-isolated Three Phase AC–DC Converter

2.1 Introduction

This chapter presents a novel matrix based non-isolated three phase AC–DC converter. The proposed converter topology provides large step down voltage gain than a traditional three phase PWM buck rectifier without compromising input power quality and power conversion efficiency. Section 2.2 presents the brief review of three phase non-isolated buck rectifiers. In Sect. 2.3, the new contributions of this chapter are discussed. Section 2.4 describes the topology, modulation scheme and principle of operation of the proposed converter in details. Steady state analysis and design are presented in Sect. 2.5 which includes voltage and current stress calculation on the semiconductor devices, accurate estimation of input and output voltage/current ripple, filter design and effect of modulation index on input current THD. In Sect. 2.6, digital simulation of the converter is carried out and results are discussed. In Sect. 2.7, small signal modeling followed by closed loop PI controller is presented for the proposed converter. A comparative evaluation of the proposed converter with the traditional six switch buck rectifier is carried out in Sect. 2.8. Subsequently, experimental results are presented in Sect. 2.9. Section 2.10 provides the conclusion.

2.2 A Brief Review of Three Phase Non-isolated Buck Rectifier

Basically, two types of topologies are possible for AC–DC conversion: (1) boost type rectifier (2) buck type rectifier. For applications, where lower output DC voltage is required, buck type of rectifiers are used since the boost type structure has limitation over minimum output DC voltage. Moreover, buck type topologies have other benefits over boost type including direct start up and the over current protection in case of

© Springer Nature Singapore Pte Ltd. 2018
A. K. Singh, *Analysis and Design of Power Converter Topologies*
for Application in Future More Electric Aircraft, Springer Theses,
https://doi.org/10.1007/978-981-10-8213-9_2

an output short circuit [1]. Therefore, the buck type rectifiers are of high interest for applications such as telecommunication, power supplies for process technology and More Electric Aircraft (MEA) systems [2–4].

The three phase buck rectifier with six switches is conventional topology and has been widely discussed in the literature [5–9]. In [10], an AC/DC matrix converter with an optimized modulation strategy is presented for V2G application. A three phase discontinuous mode buck rectifier has been presented in [11] where input power quality is improved by modifying the traditional control algorithm and modulation scheme.The comprehensive design of a thee phase three-switch buck type PWM rectifier is carried out in [1].

The zero current switched buck rectifier circuits are presented in [12, 13] which use an axillary circuit to achieve zero voltage switching with unity power factor. The output of the buck rectifier is controlled by varying the modulation index, m (also known as modulation depth) which essentially controls the pulse width of the gating signal to the switching devices. In theory, the output voltage of a buck rectifier can be varied from $[\frac{3}{2}V_m]$ to 0 by varying the modulation index, m from 1 to 0 where, V_m is the peak value of input phase voltage. However, if the desired regulated output DC voltage is lower than $[\frac{3}{4}V_m]$, the converter always has to be operated at the modulation index, $m \leq 0.5$ which under-utilizes the converter capability. Moreover, operation of the converter at lower value of m increases the THD of input phase currents and the switch rms current contributing to reduced input power quality and increased switch conduction loss, respectively.

For large step down voltage gain, the matrix based AC–DC converter with an isolation transformer can be used [13–15]. By changing the turns-ratio of the transformer, the desired output DC voltage can be obtained. However, the applications where electrical isolation is not mandatory, the use of transformer reduces the power density with additional weight and power loss. Moreover, the leakage inductance of the transformer presents additional current commutation problems resulting in increased switch voltage stress and duty cycle loss [14, 15].

2.3 New Contributions of the Chapter

In this chapter, a non-isolated matrix based three phase AC–DC rectifier is proposed which provides half of the voltage gain achieved by the conventional three phase buck type rectifiers without compromising the input power quality and the power conversion efficiency. The block diagram of the proposed converter is shown in Fig. 2.1. The input three phase AC voltage is first filtered using an input LC filter followed by matrix (3 × 1) topology. The matrix topology converts the three phase line frequency AC voltages into an intermediate high frequency AC voltage. The high frequency AC voltage is then fed to a Current Doubler Rectifier (CDR) circuit [16, 17] which in turn, rectifies it to output DC voltage. The intermediate high frequency AC voltage generated by the matrix topology allows the use of CDR circuit which essentially reduces the output DC voltage by half. Moreover, the high frequency AC

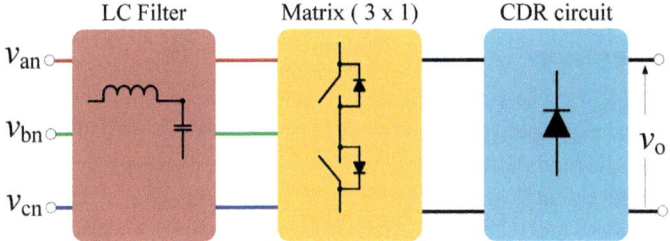

Fig. 2.1 Block diagram of the proposed matrix based AC–DC converter

voltage reduces the passive filter elements of (L and C) of the current doubler. A SVM based modulation scheme is proposed for the matrix converter for improved input power quality.

In summary, the novelty and contributions of the chapter are as follows:

- A *novel* matrix converter (3 × 1) based topology is proposed for large step down non-isolated buck rectification. Subsequently, a modified SVM based modulation scheme is presented which provides simpler implementation and improves overall power conversion efficiency;
- the proposed SVM based modulation scheme requires single control for each of the matrix switches unlike the proposed two independent control for each matrix switch in [18] and therefore, does not need switch body diode conduction and therefore, facilitates,
  [1] reduced number of isolated gate drivers (six for six matrix switches)
  [2] no body diode loss (conduction loss and reverse recovery loss). Moreover, the switching sequence is arranged symmetrically to provide symmetrical bipolar high frequency AC voltage at the matrix output unlike [15];
- comprehensive steady state analysis and design equations are presented based on the modes of operation which are further validated using both simulation and experimental tests;
- detailed mathematical derivations are carried out for accurate input/output voltage/current ripple for the proposed matrix converter for the proposed modulation scheme;
- small signal modelling followed by closed loop PI controller is designed for the proposed converter which is subsequently, validated through digital simulation.
- the comprehensive loss analysis of the proposed converter is carried out which is subsequently verified by experimental test results; and
- a comparative evaluation of the proposed converter with the traditional six switch buck rectifier [19] is presented to demonstrate the benefits of the proposed converter.

2.4   Topology, Modulation Scheme and Principle of Operation

In this section, the topology of matrix based non-isolated converter is discussed. Subsequently, the modulation scheme for the proposed converter is derived based on SVM. The different modes of operation of the converter is discussed in details which is used for analysis and design in Sect. 2.5. Figure 2.2 shows the circuit schematic of the proposed matrix based three phase AC–DC converter. It consists of three-phase ac input (v_{an}, v_{bn}, v_{cn}), input filter with inductors (L_a, L_b, L_c) and capacitor (C_a, C_b, C_c), six bidirectional switches (S_1-S_6), two diodes, D_1 and D_2, two output filter inductors, L_{f1} and L_{f2}, output capacitor, C_o and load resistor, R_o.

2.4.1   Assumptions

The following assumptions have been taken for deriving the switching scheme of the proposed AC–DC converter (Fig. 2.2).

- The filter capacitor voltages are of purely sinusoidal in shape and in phase with the three phase input AC voltages which are given by;

$$v_{an} = V_m sin(\theta); \; v_{bn} = V_m sin\left(\theta - \frac{2\pi}{3}\right);$$

$$v_{cn} = V_m sin\left(\theta + \frac{2\pi}{3}\right)$$

(2.1)

where, V_m is the peak of the three phase input voltages and θ is in radians;

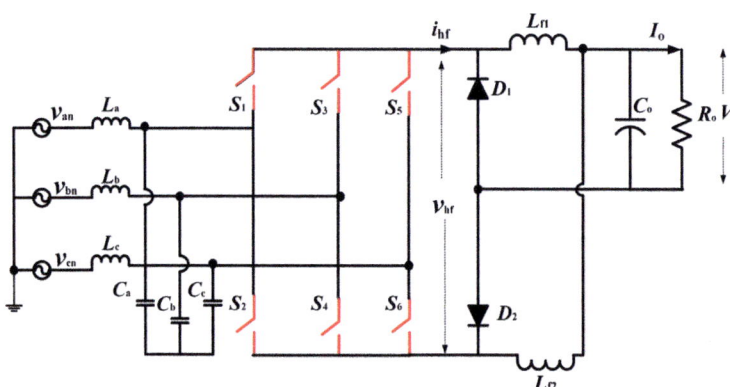

Fig. 2.2 Circuit schematic of the proposed three phase AC–DC rectifier. Each matrix switch is formed by connecting two back to back MOSFETs

- reactive current due to the filter capacitors are neglected;
- the output DC current I_o is assumed to be constant and ripple free. Consequently, the current in each of the filter inductors, L_{f1} and L_{f2} is assumed to be $\frac{I_0}{2}$ and
- the dead time, t_d between the two adjacent matrix switches which is required to avoid short circuiting of the input filter capacitors is assumed to be zero.

2.4.2   Modulation Scheme

To derive the switching signals for the matrix converter, the three-phase input AC voltages are divided into six equal sectors. During each sector, the current vector, I_{ref} can be synthesized by SVM method. For sector-1, I_{ref} can be written as,

$$I_{ref} T_s = i_{ab} t_\alpha + i_{ac} t_\beta \tag{2.2}$$

where, t_α and t_β are the time for which i_{ab} and i_{ac} flow through the circuit, respectively for one switching cycle, respectively. If T_s is the time period of one switching cycle, then the duration t_α and t_β is derived as

$$t_\alpha = m T_s \sin\left(\frac{2\pi}{3} - \theta\right) \tag{2.3}$$

$$t_\beta = m T_s \sin\left(\theta - \frac{\pi}{3}\right) \tag{2.4}$$

$$t_0 = T_s - t_\alpha - t_\beta \tag{2.5}$$

For sector-1, θ varies from $\frac{\pi}{3}$ radian to $\frac{2\pi}{3}$ radian as shown in Fig. 2.3. The time interval, t_o represents the zero period for the matrix converter. During this interval, the output voltage of the matrix converter remains zero. Once the duration, t_α, t_β and t_o are calculated, the next step is to arrange the time durations in a particular sequence to generate symmetrical bipolar high frequency AC output voltage. In the proposed modulation scheme, the switching period, T_s is divided into two equal parts. In the first half, the positive voltage is generated whereas in the second half, the negative voltage is generated. However, the sum of the each time duration, t_α, t_β and t_0 are distributed in such a way that they remain unchanged and satisfy (2.2). Each of the matrix switches, $S_1 - S_6$ is realized by connecting two back to back MOSFETs. This particular arrangement of the MOSFETs in the matrix switch facilitates four quadrant operation. In the proposed modulation scheme, the two back to back connected MOSFETs are controlled with a single control signal which is different from the traditional SVM based modulation scheme where both MOSFETs are independently controlled. The single control for each of the matrix switch requires only six isolated gate drivers for the matrix operation. When control signal is ON, both of the MOSFETs are ON and

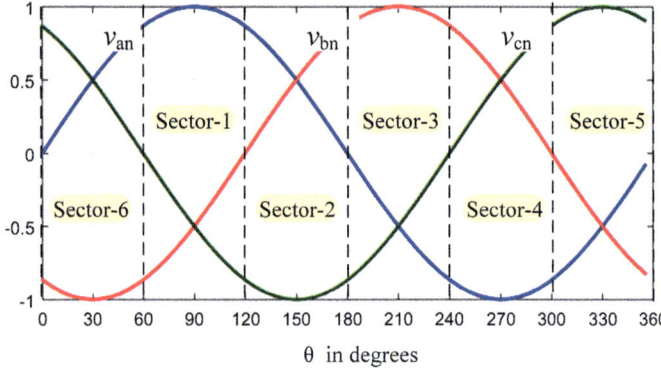

Fig. 2.3 Operation of the converter is divided into six similar sectors. Sector-1 ranges from $\theta = 60-120°$

therefore, the current always flow through the MOSFET's channel whereas when control signal is OFF, both of the MOSFETs are in OFF state and no current flows through the matrix switch. The four state of the switch in the proposed switching scheme is shown in Fig. 2.4. Figure 2.4a, b show the state of the matrix switch, S_1 when it is in ON state. The current flows through the channel of the MOSFETs in both condition. Figure 2.4c, d show the state of the matrix switch when it is in OFF state. No current flows through the switch, S_1. However, one of the body diodes of the MOSFETs is forward biased due to the voltage across the switch S_1. As shown in Fig. 2.4c, the body diode, D_{sw2} is forward biased whereas in Fig. 2.4d, the body diode, D_{sw1} is forward biased.

It is worth mentioning that there is no state of the matrix switch where body diodes of the MOSFETs conduct. Therefore, in the proposed modulation scheme, losses due to body diodes are eliminated. Moreover, as one of the body diodes of the matrix switch remains forward biased in OFF state, it turn ON with Zero Voltage resulting

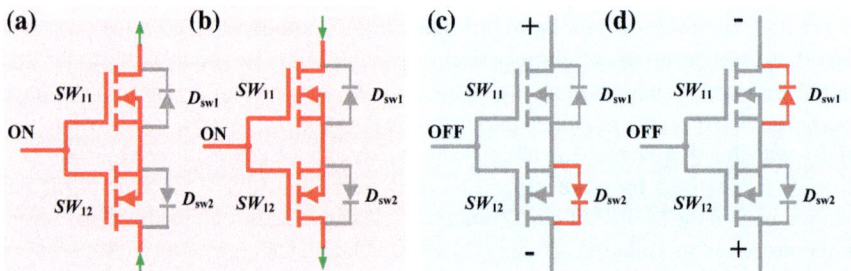

Fig. 2.4 The different states of the matrix switch, S_1. **a** Switch S_1 is ON and current is flowing in the +ve direction. **b** Switch S_1 is ON and current is flowing in the −ve direction. **c** Switch S_1 is OFF and +ve voltage appears across the switch. **d** Switch S_1 is OFF and −ve voltage appears across the switch

in Zero Voltage Switching (ZVS) of the MOSFET. For example, in Fig. 2.4c, the MOSFET, SW_{12} turns ON with ZVS. Therefore, the proposed switching scheme reduces both the switch conduction loss and the switching loss.

2.4.3 Modes of Operation

The complete modes of operation of the proposed converter can be divided into 6 modes. For the first 3 modes (mode-1 to mode-3), the inversion signal $U = 1$, whereas for next 3 modes (mode-4 to mode-6), the inversion signal $U = 0$.

Mode-1 ($t_0 \leq t \leq t_1$): During this mode, the inductor currents, i_{Lf1} and i_{Lf2} flows through the diodes, D_1 and D_2 respectively resulting in zero voltage ($v_{hf} = 0$) across the matrix converter output. The output current, I_o is shared equally in both diodes, D_1 and D_2. No power transfer happens during this mode. The governing equation during this mode of operation is given as,

$$v_{hf}(t - t_o) = 0; \; i_{hf}(t - t_o) = 0; \; i_{D1}(t - t_o) = i_{D2}(t - t_o) = \frac{I_o}{2}; \quad (2.6)$$

Mode-2 ($t_1 \leq t \leq t_2$): During this mode of operation, switches S_1 and S_4 are turned ON resulting in v_{ab} at the matrix output. The diode, D_1 remains off during this mode. The diode, D_2 shares all the output current. The governing equation during this mode of operation is given as,

$$v_{hf}(t - t_1) = v_{ab}; \; i_{hf}(t - t_1) = \frac{I_o}{2}; \; i_{D1}(t - t_1) = 0; \; i_{D2}(t - t_1) = I_o; \quad (2.7)$$

Mode-3 ($t_2 \leq t \leq t_3$): During this mode of operation, switches S_1 and S_6 are turned ON resulting in v_{ac} at the matrix output. The diodes, D_1 remains off during this mode. The diode, D_2 shares all the output current. The governing equation during this mode of operation is given as,

$$v_{hf}(t - t_2) = v_{ac}; \; i_{hf}(t - t_2) = \frac{I_o}{2}; \; i_{D1}(t - t_2) = 0; \; i_{D2}(t - t_2) = I_o; \quad (2.8)$$

Mode-4 ($t_3 \leq t \leq t_4$): This mode is exactly similar to mode-1. During this mode, the inductor currents, i_{Lf1} and i_{Lf2} flows through the diodes, D_1 and D_2 respectively resulting in zero voltage ($v_{hf} = 0$) across the matrix converter output. The output current, I_o is shared equally in both diodes, D_1 and D_2. No power transfer happens during this mode. The governing equation during this mode of operation is given as,

$$v_{hf}(t - t_3) = 0; \; i_{hf}(t - t_3) = 0; \; i_{D1}(t - t_3) = i_{D2}(t - t_3) = \frac{I_o}{2}; \quad (2.9)$$

Mode-5 $(t_4 \leq t \leq t_5)$: During this mode of operation, switches S_2 and S_3 are turned ON resulting in $-v_{ab}$ at the matrix output. The diodes, D_2 remains off during this mode. The diode, D_1 shares all the output current. The governing equation during this mode of operation is given as,

$$v_{hf}(t - t_4) = -v_{ab}; i_{hf}(t - t_4) = \frac{I_o}{2}; i_{D1}(t - t_4) = I_o; i_{D2}(t - t_4) = 0; \quad (2.10)$$

Mode-6 $(t_5 \leq t \leq t_6)$: During this mode of operation, switches S_2 and S_5 are turned ON resulting in $-v_{ac}$ at the matrix output. The diodes, D_2 remains off during this mode. The diode, D_1 shares all the output current. The governing equation during this mode of operation is given as,

$$v_{hf}(t - t_5) = -v_{ac}; i_{hf}(t - t_5) = \frac{I_o}{2}; i_{D1}(t - t_5) = I_o; i_{D2}(t - t_5) = I_o; \quad (2.11)$$

The end of mode-6 completes the operation of the proposed converter for one switching cycle. A mode similar to mode-1 starts after the end of mode-6 and thus, generates symmetrical bipolar high frequency AC at the matrix output.

2.5 Steady State Analysis and Design

In this section, the steady state analysis of the converter is carried out based on the modes of operation as described in Sect. 2.4. Subsequently, design equations such as voltage and current stresses are derived for the proposed converter.

2.5.1 Voltage Gain of the Converter

The derivation of voltage gain is based on the assumption that the output current, I_o is constant and ripple free through out the switching cycle (T_s). By volt-time balance across one of the filter inductors, voltage gain of the converter can be derived. Based on the modes of operation shown in Fig. 2.5, the voltage across inductor is shown in Table 2.1:

The volt-time balance across inductor, L_{f1} results in,

$$-V_o\frac{t_o}{2} + (v_{ab} - V_o)\frac{t_\alpha}{2} + (v_{ac} - V_o)\frac{t_\beta}{2} - V_o\left(\frac{T_s}{2}\right) = 0 \quad (2.12)$$

where, T_s is defined as,

$$T_s = t_o + t_\alpha + t_\beta \quad (2.13)$$

Fig. 2.5 Theoretical modes
of operation of the proposed
AC–DC converter.
a Reference sawtooth signal,
ST for switching signal
generation. **b** Inversion
signal, U. **c** Switching signal
sequence based on SVM
method. **d** High frequency
AC output, v_{hf}. **e** High
frequency AC current, i_{hf}.
f Current in diode, D_1.
g Current in diode, D_2

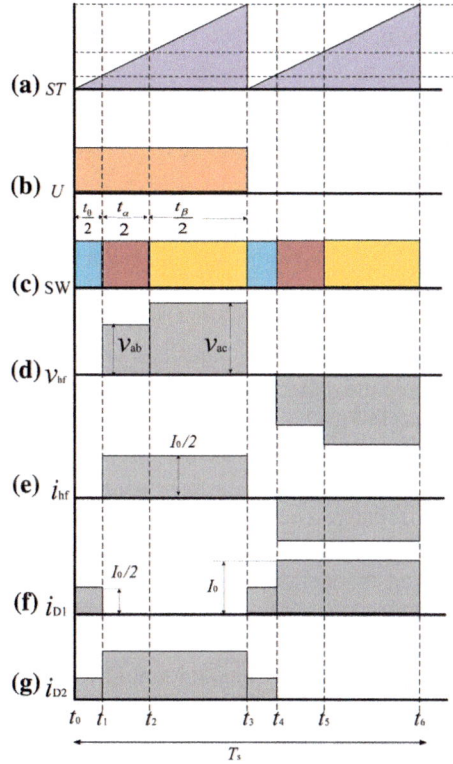

Table 2.1 Voltage across
inductor, L_{f1}

Duration	Voltage across inductor, L_{f1}
$t_0 - t_1$	$-V_o$
$t_1 - t_2$	$v_{ab} - V_o$
$t_2 - t_3$	$v_{ac} - V_o$
$t_3 - t_4$	$-V_o$
$t_4 - t_5$	$-V_o$
$t_5 - t_6$	$-V_o$

Simplifying (2.12) and (2.13) result in,

$$2V_o T_s = v_{ab}t_\alpha + v_{ac}t_\beta \tag{2.14}$$

From Eq. (2.1)–(2.5) the above equation can be further simplified to,

$$V_o = \frac{3}{4}m V_m \tag{2.15}$$

where, V_m is the peak voltage of the input AC phase voltage. By controlling the value of the modulation index, m, the output voltage, V_o is controlled.

2.5.2 Voltage Stresses

In this sub-section, the voltage stresses on the active and passive devices are determined. The maximum line to line input voltage, $V_{ll,max}$ is,

$$V_{ll,max} = \sqrt{3}V_m \tag{2.16}$$

Each of the matrix switches is realized by connecting two MOSFETs back to back as shown in Fig. 2.4. It is to be noted that maximum voltage across each of the switches is $\sqrt{3}V_m$. The voltage across one of the two switches of the matrix switch is zero as it is forward biased.

Similarly, the maximum voltage stress across diodes, D_1 and D_2 is the maximum value of high frequency AC output of the matrix converter, v_{hf} which is $\sqrt{3}V_m$. During mode-1 and mode-4 the voltage across both the diodes is zero (assuming zero forward voltage drop across diodes).

The filter inductors, L_{f1} and L_{f2} and filter capacitor, C_o are selected for output voltage V_o. A sufficient amount of margin should be considered while selecting the active and passive devices.

2.5.3 Current Stresses

The calculation of current stress is based on the assumption that the load current, I_o is constant and ripple free and switching frequency, f_s is much higher than the input voltage frequency, f_i. The current stress in each of the semiconductor devices is evaluated as follows.

2.5.3.1 Current Stresses in the Matrix Switches

The average and *rms* values of current in the matrix switch are calculated which are critical for selecting the suitable switch for a given specification. Both average and *rms* values are calculated for half of the mains period ($\theta = 0$ to π). For a given load current, I_o the amplitude of current in the matrix switches is $\frac{I_o}{2}$. The calculation of the average switch current is two step process. In the first step, the average is calculated for a switching cycle whereas in the second step, it is calculated for entire line period (2π radian). During half of the line period (π), the current in the switches can be divided into three parts. For 0 to $\frac{\pi}{3}$, the current $\frac{I_o}{2}$ flows for only $\frac{t_a}{2}$ duration. For $\frac{\pi}{3}$

to $\frac{2\pi}{3}$ duration, current $\frac{I_o}{2}$ flows for $\frac{t_\alpha + t_\beta}{2}$ duration. For the rest, For $\frac{2\pi}{3}$ to π duration he current $\frac{I_o}{2}$ flows for only $\frac{t_\beta}{2}$ duration.

Based on the above discussion and using (2.3) and (2.4), the average current in the switches is given by,

$$i_{SW,avg} = \frac{mI_o}{4\pi}\left[\int_0^{\frac{\pi}{3}} \sin(\theta)\,d\theta + \int_{\frac{\pi}{3}}^{\frac{2\pi}{3}}\left(\sin\left(\frac{2\pi}{3} - \theta\right) + \sin\left(\theta + \frac{\pi}{3}\right)\right)d\theta + \int_{\frac{2\pi}{3}}^{\pi} \sin(\pi - \theta)\,d\theta\right] \tag{2.17}$$

which is simplified to

$$i_{SW,avg} = \frac{I_o}{2\pi}m \tag{2.18}$$

Similarly, the rms current in the switches can be calculated which is given by the equation,

$$i_{SW,rms} = \sqrt{\frac{1}{\pi}\frac{mI_o^2}{8}\left(\int_0^{\frac{\pi}{3}} \sin(\theta)d\theta + \int_{\frac{\pi}{3}}^{\frac{2\pi}{3}} \sin\left(\frac{2\pi}{3} - \theta\right)d\theta + H\right)} \tag{2.19}$$

where, $H = \int_{\frac{\pi}{3}}^{\frac{2\pi}{3}} \sin(\theta - \frac{\pi}{3})d\theta + \int_{\frac{2\pi}{3}}^{\pi} \sin(\pi - \theta)d\theta$
Simplifying the (2.19) results in,

$$i_{SW,rms} = \frac{I_o}{2}\sqrt{\frac{m}{\pi}} \tag{2.20}$$

It is important to note here that the use of current doubler rectifier circuit in the output side reduces the current amplitude in the switches by half. As conduction loss in the switches is directly proportional to the switch *rms* current, there is significant reduction in conduction loss for the matrix switches.

2.5.3.2   Form Factor of Matrix Switch Current

The form factor is defined as the ratio of *rms* current and average current. The lower form factor of current indicated lower loss for a given output power. The form factor, k_f for the matrix switch current is given by,

$$k_f = \sqrt{\frac{\pi}{m}} \tag{2.21}$$

Fig. 2.6 Form factor of the matrix switch current

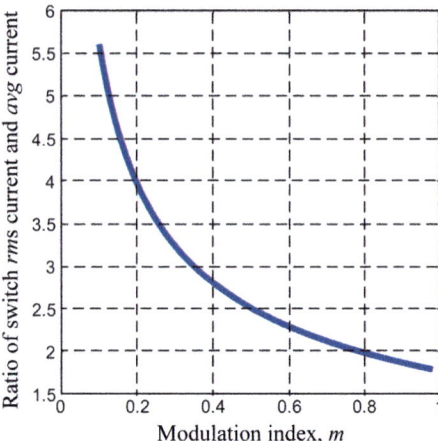

Table 2.2 Current through the diodes, D_1 and D_2

Duration	Current in diode, D_1, D_2
$t_0 - t_1$	$\frac{I_o}{2}$
$t_1 - t_2$	I_o
$t_2 - t_3$	I_o
$t_3 - t_4$	$\frac{I_o}{2}$
$t_4 - t_5$	0
$t_5 - t_6$	0

Figure 2.6 shows that for a given average current, switch *rms* current increases with lower modulation index which contributes to more switch losses. Therefore, it is not preferred to design the operating point of the converter at lower modulation index.

2.5.3.3 Current Stress in the Diodes

The current stresses in the current doubler rectifier diodes are calculated based on the modes of operation as shown in Fig. 2.4. Table 2.2 shows the current in diode during one switching cycle. The average current, $i_{D_1,avg}$ is calculated for the complete switching cycle and is found to be $\frac{I_o}{2}$. Similarly the rms value, $i_{D_1,rms}$ is calculated. Similarly, *avg* and *rms* currents in diode, D_2 can be evaluated which is the same as diode, D_1.

$$i_{D_1,avg} = \frac{I_o}{2} \tag{2.22}$$

$$i_{D_1,rms} = \frac{I_o}{\sqrt{2}} \tag{2.23}$$

2.5.4 Estimation of the Input and Output Voltage/Current Ripple for the Proposed Matrix Based AC–DC Converter

In this section, the accurate method of estimating the input and output voltage/current ripple for matrix based AC–DC converter is carried out. Subsequently, the design equation has been derived. To derive voltage and current ripple, following assumptions are made:

- all the input filter inductors, $L_a = L_b = L_c = L_i$;
- all the input filter capacitors $C_a = C_b = C_c = C_i$; and
- the input displacement power factor is unity.

To calculate the voltage/current ripple in the input side, first the input voltage ripple is calculated. Subsequently, the input current ripple is calculated based on the input voltage ripple. This section is divided into two parts. The first part deals with input voltage/current ripple whereas second deals with output voltage/current ripple.

2.5.4.1 Input Voltage and Current Ripple

Assuming the frequency of the high frequency AC output of the matrix converter is $f_s = \frac{1}{T_s}$, the output voltage ripple of the input filter capacitor is derived. During zero period, $\frac{t_0}{2}$ the input current charges the filter capacitor which results in peak to peak voltage ripple, $v_{C,pk}$. The voltage current relationship of the capacitor is given by,

$$i(t) = C_i \frac{dv_c(t)}{dt} \qquad (2.24)$$

where, $v_c(t)$ is voltage across the input filter capacitor.

From (2.24), the peak to peak voltage ripple can be derived by,

$$\Delta v_c(t) = \frac{1}{C_i} i(t) dt \qquad (2.25)$$

Assuming unity power factor, the phase -a current can be written as, $i(t) = I_m \sin \theta$. where, $\theta = 2\pi f_i t$. f_i is the input line frequency. Based on the switching cycle shown in Fig. 2.7, the unfiltered input current is shown in Fig. 2.8. It is to be noted that when no current is flowing, the input filter inductor current is charging the input filter capacitor resulting in peak to peak input voltage ripple. The time period for which the current, $i(t)$ charges the filter capacitor, C_i during half of the switching cycle is derived as,

$$t_c = \frac{1}{2f_s}(1 - m \sin \theta) \qquad (2.26)$$

From (2.25) and (2.26),

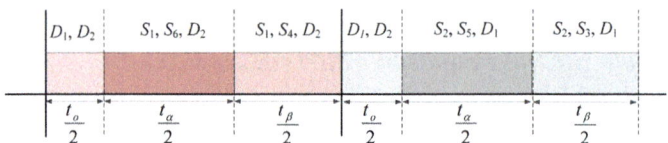

Fig. 2.7 Switching sequence for sector-1 during a switching cycle, T_s

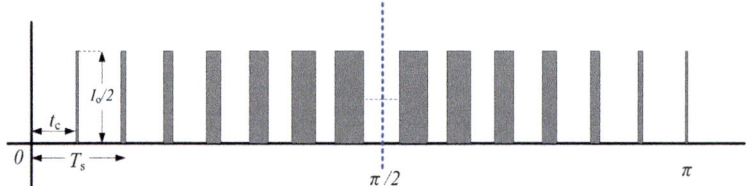

Fig. 2.8 Unfiltered input phase-a current

$$\Delta v_c(\theta) = \frac{I_m \sin \theta}{2C_i f_s}(1 - m \sin \theta) \tag{2.27}$$

For a load current, I_o, the peak of the input current I_m is found by input-output power balance, which is given as,

$$I_m = \frac{m I_o}{2} \tag{2.28}$$

From (2.27) and (2.28), the input filter capacitor ripple is derived as,

$$\Delta v_c(\theta) = \frac{I_o}{4C_i f_s} m \sin \theta (1 - m \sin \theta) \tag{2.29}$$

From (2.29), following observation can be made:

- for $\theta = 0$ and $\theta = \pi$, the voltage ripple is zero; and
- for $\theta = \frac{\pi}{2}$ and $m = 1$, voltage ripple is zero.

Lets define, $\alpha = \frac{I_o}{4C_i f_s}$. A graph is plotted for $\frac{\Delta V_c}{\alpha}$ with θ at different modulation index to show the variation of input capacitor voltage ripple as shown in Fig. 2.9. The maximum voltage ripple from (2.29) is found where $m \sin \theta = \frac{1}{2}$ which provides,

$$\Delta V_{c,max} = \frac{I_o}{16C_i f_s} \tag{2.30}$$

The angle θ, at which the maximum capacitor voltage ripple occurs is given by,

$$\theta_{max} = \sin^{-1}\left(\frac{1}{2m}\right) \tag{2.31}$$

Fig. 2.9 Variation of input capacitor voltage ripple

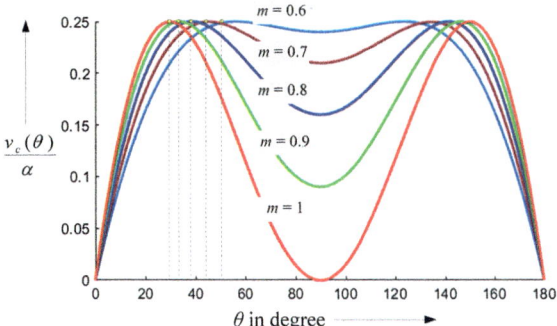

where, $m \geq 0.5$

The input current ripple can be given in terms of capacitor voltage ripple as,

$$\Delta i(\theta) = \frac{\Delta v_c(\theta)}{4\pi f_s L_i} \tag{2.32}$$

From (2.30) and (2.32), the maximum value of current ripple in the line inductor of value, L_i is given by,

$$\Delta I_{max} = \frac{1}{64\pi} \frac{I_o}{C_i L_i f_s^2} \tag{2.33}$$

It should be noted that by increasing the switching frequency, f_s and line inductance, L_i, the inductor current ripple can be minimized.

2.5.4.2 Output Voltage and Current Ripple

In this section, the output voltage and current ripple are calculated. As shown in Fig. 2.2, the inductors, L_{f1} and L_{f2} are identical output inductors. The current ripple in one inductor is shifted by 180° with respect to another inductor. Assuming the peak to peak current ripple in both the inductors are triangular and identical, the output current ripple and the output voltage ripple are calculated. This section is divided into two subsections. In the first subsection, the current ripple in one of the output inductors is derived. Subsequently, the output ripple of the total load current is estimated. In the second subsection, the output voltage ripple is calculated.

Assume, $L_{f1} = L_{f2} = L_o$. As shown in Fig. 2.5f, the diode D_1 conducts for $\frac{T_s + t_o}{2}$ duration for which the voltage across the inductor, L_{f1} remains $(-V_o)$. The current in inductor, L_{f1} falls with a constant slope which results in peak to peak current ripple. From Fig. 2.10, the output current ripple in the inductor L_{f1} is given by,

$$\Delta I_{Lf1} = \frac{V_o}{2L_o}\left(T_s + t_o\right) \tag{2.34}$$

Fig. 2.10 Output inductor current ripple, i_{Lf1} and i_{Lf2}

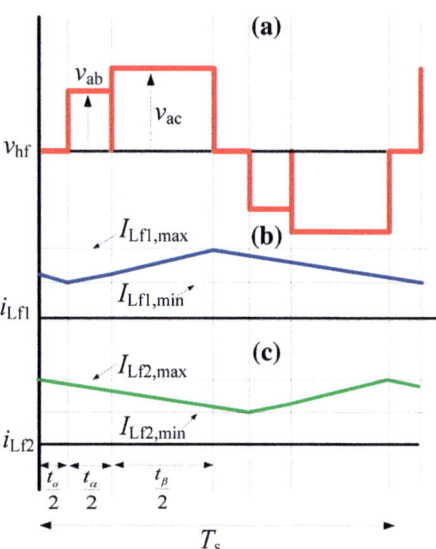

Simplifying the above equation gives,

$$\Delta I_{LF1} = \frac{V_o}{L_o f_s}\left(1 - \frac{m}{2}\sin\theta\right) \tag{2.35}$$

Assuming ripple to be of triangular shape, the output current ripple is half in the magnitude and double in frequency due to the interleaving of the output filter inductors which gives,

$$\Delta I_{Lo} = \frac{V_o}{2L_o f_s}\left(1 - \frac{m}{2}\sin\theta\right) \tag{2.36}$$

As shown in (2.36), the output current ripple is not constant throughout the sector. It varies with the angle, θ. Assuming, $\beta = \frac{V_o}{2L_o f_s}$, the peak to peak inductor current ripple with angle θ can be shown for different modulation index, m (Fig. 2.11).

The maximum value of output current ripple is at the starting and the ending of sector. Thus, the maximum value of current ripple is given by,

$$\Delta I_{Lo,MAX} = \frac{V_o}{2L_o f_s}\left(1 - \frac{\sqrt{3}m}{4}\right) \tag{2.37}$$

From (2.37), it is evident that the maximum value of current ripple depends on the modulation index, filter inductance and switching frequency. By increasing the filter inductance and switching frequency, $\Delta I_{Lo,MAX}$ can be reduced. With increase in the modulation index, m, the output current ripple, ΔI_{Lo} reduces.

Fig. 2.11 Variation of output current ripple for sector-1 ($\theta = 60-120°$)

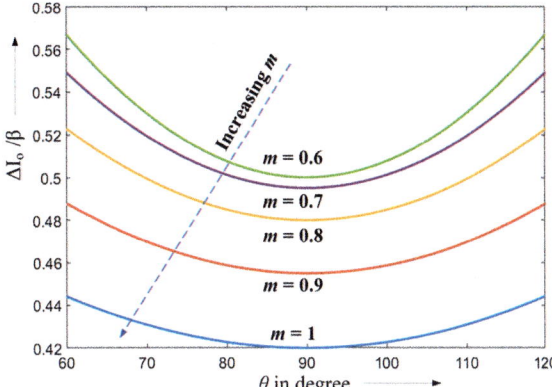

Fig. 2.12 **a** Output DC current, I_{Lo}. **b** Output DC voltage, V_o

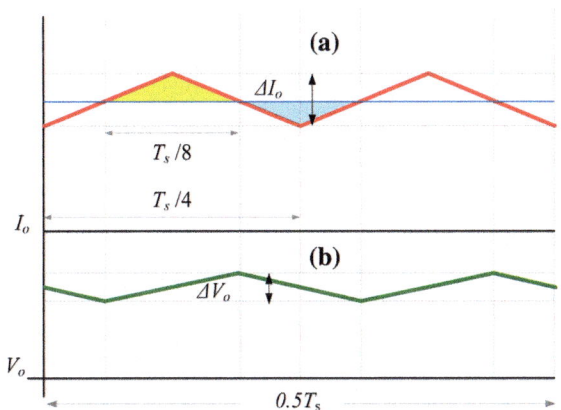

Figure 2.12 shows the output current ripple and output voltage ripple waveforms. The output capacitor is charged for $\frac{T_s}{8}$ time period and then discharges for $\frac{T_s}{8}$ as shown in Fig. 2.12b. The resulting capacitor voltage ripple is shown in Fig. 2.12c. It is to be noted that the derivation is carried for maximum output current ripple, $\Delta I_{Lo,MAX}$. The derivation of maximum capacitor voltage ripple is carried out as follows. From the basic charge, Q and voltage, V_o relationship of the capacitor, C_o,

$$\Delta V_o = \frac{\Delta Q}{C_o} \tag{2.38}$$

$$\Delta Q = \frac{1}{2} \frac{\Delta I_{Lo,MAX}}{2} \frac{T_s}{4} \tag{2.39}$$

From (2.37), (2.38) and (2.39), the maximum output voltage ripple is derived as,

$$\Delta V_{o,MAX} = \frac{V_o}{32L_oC_of_s^2}\left(1 - \frac{\sqrt{3}m}{4}\right) \qquad (2.40)$$

2.5.5   Filter Design

In this sub-section the design of filter is carried out. There are two type of filters (1) input LC filter and (2) output LC filter required in the proposed converter. The design of L and C components is carried out as follows:

2.5.5.1   Input Filter Design

Unlike PWM boost kind of rectifiers, the input filter of the proposed converter is an LC filter. For getting smooth sinusoidal current, high quality capacitors with low ESR, ESL and high current ratings are required. However, a large value of input capacitor results in capacitive reactive power leading to low power factor particularly in low power conditions. The reactive power supplied to the capacitor is given by,

$$Q_m = 3\omega_i C_i V_i^2 \qquad (2.41)$$

where, $\omega_i = 2\pi f_i$ It is desirable that even at low load, the displacement power factor does not go below a certain value. Based on this specification, the upper limit of input filter capacitance can be calculated.

After the selection of capacitor value, the inductor value is chosen. Typically the resonant frequency is kept higher than 20 times of the supply frequency and lower than one-third of the PWM frequency [20]. The input voltage and current ripple of the LC filter for the selected inductor and capacitor values are calculated using (2.30) and (2.33). The calculated values should be within the specified limits.

Based on [20], the damping resistor for the filter is selected. Figure 2.13 shows the per phase equivalent circuit of the proposed matrix converter. For the chosen configuration of the damped resistor for the input LC filter, the voltage and current transfer functions can be derived as,

$$G_{vd} = \frac{v_i(s)}{v_s(s)} = \frac{sC_i(R_d + sL_i)}{s^2R_dL_iC_i + sL_i + R_d} \qquad (2.42)$$

Fig. 2.13 Configuration of damped input filter for the proposed converter

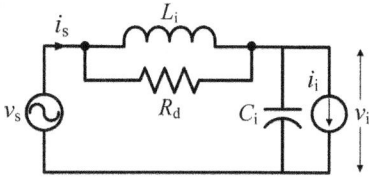

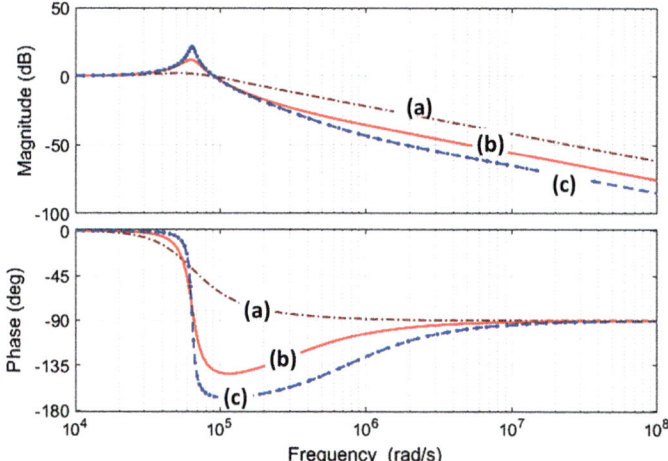

Fig. 2.14 Bode plot for the current transfer function, G_{id} for different damping resistance value.
a $R_d = 10\,\Omega$. **b** $R_d = 50\,\Omega$. **c** $R_d = 100\,\Omega$

$$G_{id} = \frac{i_i(s)}{i_s(s)} = \frac{R_d + sL_i}{s^2 R_d L_i C_i + sL_i + R_d} \tag{2.43}$$

The input source voltage, v_s is balanced and sinusoidal. However, the output current, i_i is pulse width modulated current as shown in Fig. 2.15 and therefore, it contains high frequency AC components. The high frequency AC components lead to LC oscillations. By using a damping resistor, the LC oscillations of the input filter can be attenuated. Based on the (2.42) and (2.42), the resonant frequency, $\omega_{r,i}$ and damping factor, ξ are given as,

$$\omega_{r,i} = \frac{1}{\sqrt{L_i C_i}} \tag{2.44}$$

$$\xi = \frac{1}{2R_d}\sqrt{\frac{L_i}{C_i}} \tag{2.45}$$

The frequency response plot for the current transfer function, G_{id} is plotted for different damping resistance values and shown in Fig. 2.14. The values of input inductor, L_i and the input capacitor, C_i are chosen to be $200\,\mu\text{H}$ and $1.2\,\mu\text{F}$. It is evident from Fig. 2.14 that lower the resistance value (higher the damping factor), poorer is the high frequency attenuation. Therefore, a compromise is needed between the damping factor and the filtering performance.

2.5.5.2   Output Filter Design

In the output of the proposed converter, a current doubler circuit is used which consists of two equal inductors, L_{f1} and L_{f2} and an output capacitor, C_o. The design of inductors, L_{f1} and L_{f2} is carried out in such a manner that the peak-to-peak value of the DC output inductor current ripple, $\Delta i_{Lo,Max}$ is limited to a given value. The peak to peak value of the inductor current ripple can be calculated using (2.37) With this the output inductor can be selected according to,

$$L_o \geq \frac{V_o}{2\Delta i_{Lo,MAX} f_s}\left(1 - \frac{\sqrt{3}m_{min}}{4}\right) \tag{2.46}$$

where, m_{min} is the minimum modulation index and f_s is the switching frequency of the converter. The output current is sum of the currents in inductors L_{f1} and L_{f2}. The two inductor behaves like interleaved inductors and essentially and therefore, the peak to peak current ripple in the output current, I_o is reduced. The ripple frequency of the output current, I_o becomes twice the switching frequency, f_s contributing in reducing the size of output capacitor. The output capacitor, C_o is selected in order to limit the peak-to-peak value of the output voltage ripple given by (2.40). The output capacitor value is derived as,

$$C_o \geq \frac{V_o}{32\Delta v_{o,MAX} L_o f_s^2}\left(1 - \frac{\sqrt{3}m_{min}}{4}\right) \tag{2.47}$$

2.5.6 Effect of Modulation Index (m) on the THD of the Input Line Current

In this subsection, the effect of modulation index, m on the THD of the input line current is analyzed mathematically. At lower modulation index, pulse width of the unfiltered input current becomes narrower and therefore, the THD of the current increased. In this subsection, the THD of the unfiltered input current shown in Fig. 2.15 is derived mathematically and its variation is analyzed at different modulation index, m. The analysis is subjected to the assumption that the phase voltages (which are also the voltages across the input filter capacitors) are perfectly sinusoidal. The THD of the unfiltered input current is defined by,

$$THD = \frac{\sqrt{I_{rms}^2 - I_{1,rms}^2}}{I_{1,rms}} \tag{2.48}$$

where, I_{rms} is the *rms* value of unfiltered input current and $I_{1,rms}$ is the *rms* value of the fundamental of the unfiltered input current. The *rms* value of the unfiltered

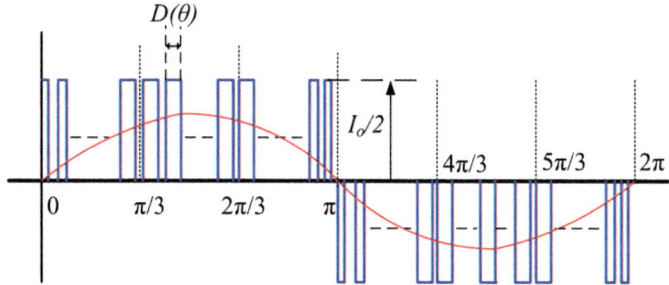

Fig. 2.15 Unfiltered input current and fundamental input current of the proposed converter

input current, I_{rms} is calculated in two steps. In the first step, the rms value of the current for a switching period, T_s is calculated. If the rms value of the current for a switching cycle is given by $I_{rms,\theta}$ then,

$$I_{rms,\theta} = \sqrt{\left(\frac{I_o}{2}\right)^2 D(\theta)} \qquad (2.49)$$

where, $D(\theta)$ is the duty cycle of the pulsed current at angle, θ. In the second step, the rms value of the current is calculated for 2π duration.

$$I_{rms} = \sqrt{\frac{1}{2\pi}\left(\int_0^{2\pi} I_{rms,\theta}^2 d\theta\right)} \qquad (2.50)$$

Simplifying the equation using (2.19) results in,

$$I_{rms} = \frac{I_o}{\sqrt{2}}\sqrt{\frac{m}{\pi}} \qquad (2.51)$$

To calculate the rms value of the fundamental current, the Fourier analysis of the unfiltered input current shown in Fig. 2.15 is carried out. The analytical harmonic solution of the unfiltered input current is given by,

$$i_a(t) = \frac{\sqrt{3}}{4}I_o m \sin(\omega_o t) +$$

$$\frac{2I_o}{\pi}\sum_{k=1}^{\infty}\sum_{l=-\infty}^{\infty}\frac{1}{k}J_n(k\frac{\pi}{2}m)\sin\left([k+l]\frac{\pi}{2}\right)\sin l\frac{\pi}{3}\sin\left(k\omega_s t + l\omega_o t\right) \qquad (2.52)$$

where, ω_s and ω_o are the switching frequency and the line frequency of the proposed converter, respectively. From (2.52), the rms value of the fundamental current, $I_{in,rms}$ is derived as,

Fig. 2.16 The variation of
the THD with modulation
index, m

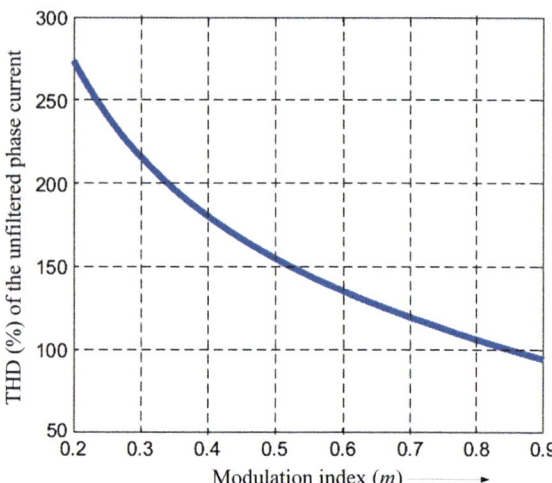

$$I_{in,rms} = \frac{\sqrt{3}}{4\sqrt{2}} m I_o \tag{2.53}$$

The THD of the unfiltered input current is calculated using (2.48), (2.51) and (2.53).

$$THD = \sqrt{\frac{16}{3m\pi} - 1} \tag{2.54}$$

Figure 2.16 shows the variation of THD with modulation index, m. It is worth noticing that the at lower modulation index, the THD of unfiltered input current increases and therefore, the converter should not be operated at lower modulation index.

2.6 Simulation of the Proposed Converter

In this section, the digital simulation of the proposed converter for the given specification in Table 2.3 is carried out. Based on the steady state analysis, the different parameters of the converter are designed. Further, the current stresses in passive and active devices and voltage and current ripple are evaluated for the proposed converter using both digital simulation and analytical calculation.

Table 2.3 Specifications of
the example converter

Parameters	Values
Input voltage, v_{abc}	115 $V_{ac}(rms)$, 400 Hz
Output voltage, V_o	90 $V\,DC$
Switching frequency, f_s	40 kHz
Output power, P_o	500 W

Table 2.4 Comparisons of the results obtained by analytic solution and digital simulation

	Parameters	Analytic solution	Digital simulation
1	Switch current (Avg)	0.65 A	0.73 A
2	Switch current (rms)	1.34 A	1.54 A
3	Diode current (Avg)	2.78 A	2.9 A
4	Diode current (rms)	3.92 A	3.93 A
5	Current ripple ($pk - pk$)	1.28 A	1.18 A
6	Voltage ripple ($pk - pk$)	0.0024 V	0.0028 V

The switching frequency of $f_s = 40\,$kHz is selected in order to achieve a good compromise between high power conversion efficiency and high power density. The specification of an example converter suitable for aircraft system is provided in Table 2.3.

With switching frequency, $f_s = 40\,$kHz the value of passive components are selected as $L_{f1} = L_{f2} = 1.2\,$mH, $C_o = 400\,\mu$F, filter inductor ($L_a = L_b = L_c$) = $200\,\mu$H, filter capacitor ($C_a = C_b = C_c$) = $1.2\,\mu$F following the design rules given in Sects. 2.5.4 and 2.5.5.

Based on the designed parameters of the converter, the converter is simulated in MATLAB 2015a for full load. The current stresses of active/passive devices as well as voltage and current ripple are evaluated which are compared with analytical solutions found by following Sects. 2.5.4 and 2.5.5. The Table 2.4 shows the comparison of the results obtained using analytic calculation and digital simulation. The results are obtained for 115 V_{rms} AC to 90 V DC conversion. The digital simulation validates the analysis of the proposed converter. The proposed matrix converter directly converts three phase line frequency AC into single phase high frequency. Figure 2.17a shows the input three phase AC, v_{an}, v_{bn}, v_{cn}. Figure 2.17b shows the high frequency AC output, v_{hf} of the matrix converter. Figure 2.17c shows the corresponding high frequency AC current, i_{hf}.

The frequency of the v_{hf} depends on the switching frequency, f_s. The switching frequency, f_s of the converter is selected as 40 kHz. Figure 2.18 shows the zoomed picture of v_{hf} and i_{hf}. A symmetrical bipolar high frequency AC of frequency ($f_s = 40\,$kHz) is generated using matrix topology. A dead time is introduced to avoid short circuiting of the input filter capacitor. The proposed modulation scheme is based on SVM modulation scheme and it provides superior input power quality. Figure 2.19 shows the three phase input current for full load (500 W). The THD of the current is estimated and found to be 3.25% as shown in Fig. 2.20. Figure 2.21 shows the input phase voltage, v_{an} and phase current, i_a for full load. The displacement power factor is found to be almost unity. The high frequency AC output of the matrix converter is processed using current doubler rectifier circuit. Figure 2.22a shows the output DC

Fig. 2.17 a Three phase
input AC voltage, v_a, v_b, v_c
[V]. **b** High frequency AC
voltage, v_{hf} [V]. **c** High
frequency AC current, i_{hf}
[A]

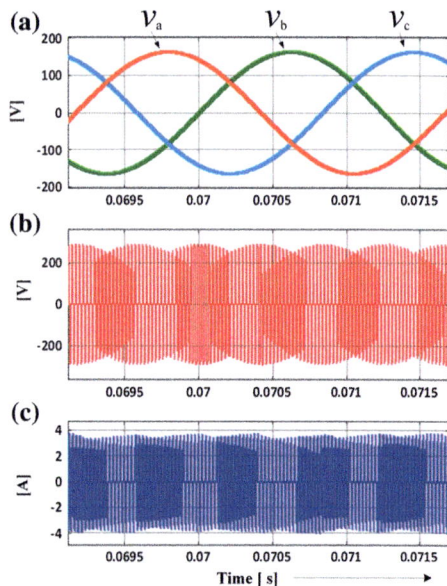

Fig. 2.18 a High frequency
AC voltage, v_{hf} [V]. **b** High
frequency AC current, i_{hf}
[A]

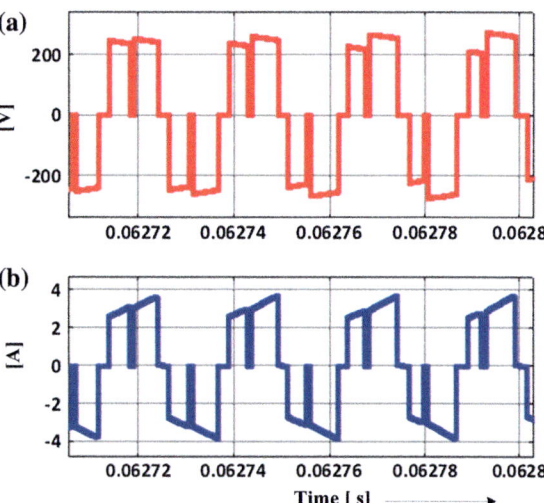

voltage, V_o. Figure 2.22b, c show the inductor current, i_{Lf1} and i_{Lf2} respectively. The
output current, I_o is sum of the two currents.

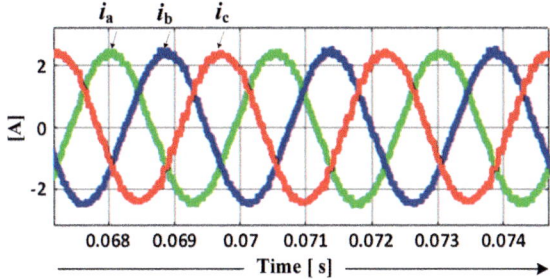

Fig. 2.19   Three phase input AC current, i_a, i_b, i_c [A]. The simulated THD of current is found to be 3%

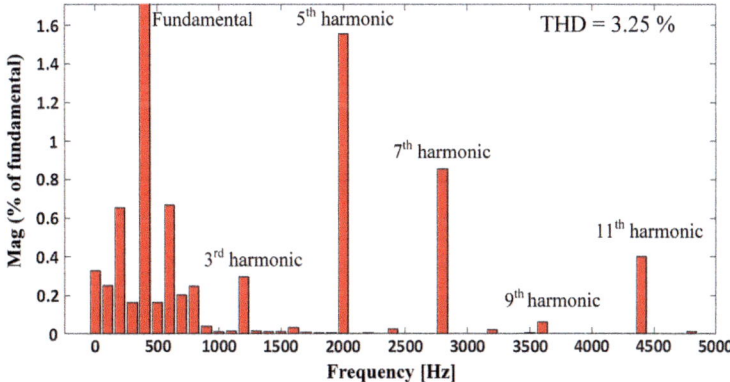

Fig. 2.20   FFT spectrum of input phase-a current. The fundamental frequency is 400 Hz. The other frequencies which have relatively larger percentage are 5th, 7th and 11th harmonics of the fundamental frequency

Fig. 2.21   Input phase-a current, i_a [A] and input phase-a voltage, v_a [V]. The displacement power factor of the converter is almost found to be unity

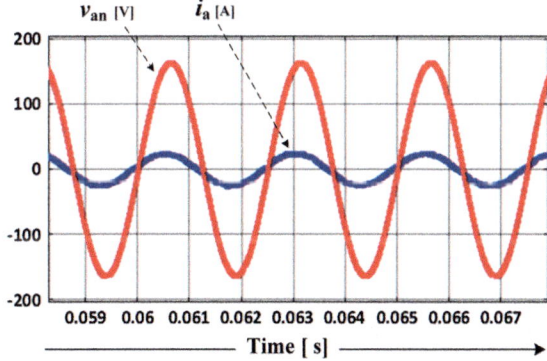

Fig. 2.22 a Output DC
voltage, V_o [v]. **b** Output
filter inductor current, i_{Lf1}
[A]. **c** Output filter inductor
current, i_{Lf2} [A]

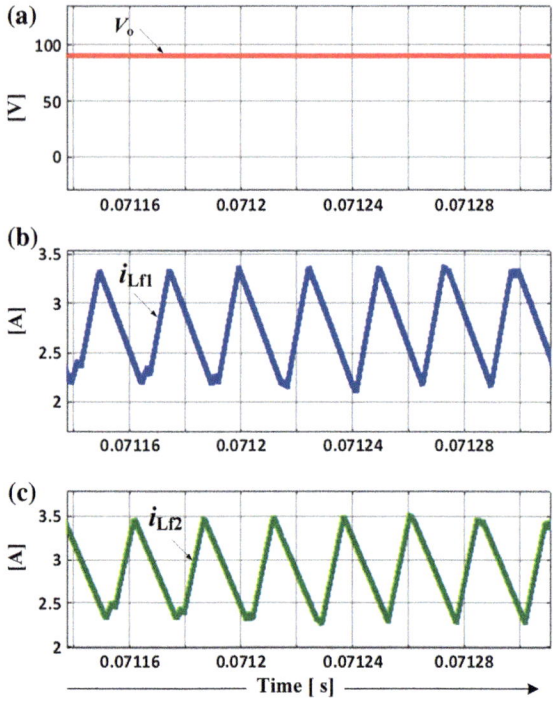

2.7 Closed Loop Control of the Proposed Converter

In this section, the closed loop control of the proposed converter is presented.
Figure 2.23 shows the block diagram for the closed loop control of the proposed
converter. The three phase input voltage, output voltage and output current is sensed
and given to controller for generating switching signals for the matrix switches. This
section is divided into three subsections. In the first subsection, the small signal modelling of the converter is carried out and subsequently, the plant transfer functions
are obtained. In second subsection, a two loop PI control consisting of an outer slow
voltage loop and an inner fast current loop, is designed and its performance is analysed. In third subsection, the proposed converter is simulated in MATLAB Simulink
with designed controller and its performance is validated for step load change from
$15-100\%$.

2.7.1 Small Signal Modelling of the Proposed Converter

Figure 2.24 shows the equivalent circuit model of the matrix based AC–DC converter.
The input voltage is symmetrical bipolar high frequency voltage indicated by $v_{hf}(t)$.

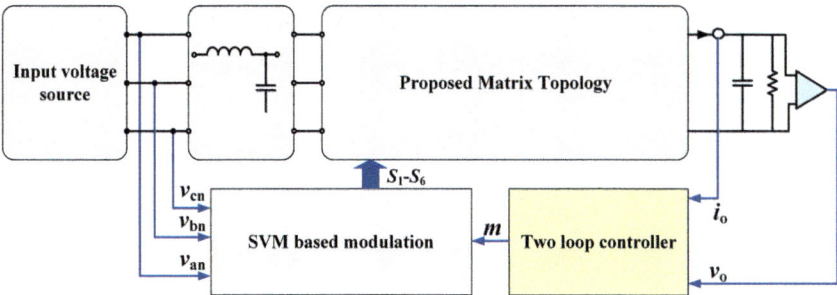

Fig. 2.23 Block diagram of the close loop control of the proposed converter

Fig. 2.24 Equivalent circuit model of three phase matrix based AC–DC converter

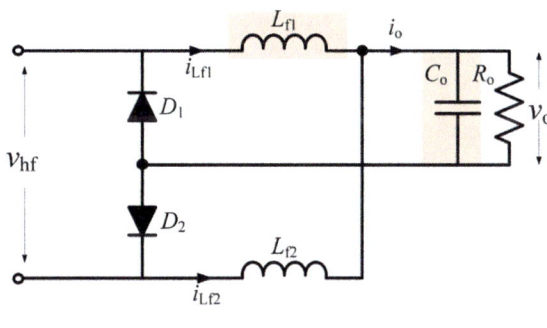

Fig. 2.25 Voltage across the output filter inductor, L_{f1} during one switching cycle, T_s in sector-1

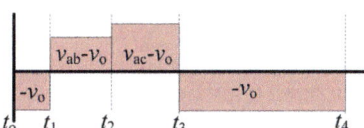

The input voltage, $v_{hf}(t)$ is synthesized by two of the three line-to-line input voltage vectors based on the sector of operation. The three state variables in the circuit are output filter inductor currents, $i_{Lf1}(t)$, $i_{Lf2}(t)$ and the output filter capacitor voltage, $v_c(t)$. The inductor currents $i_{Lf1}(t)$ and $i_{Lf2}(t)$ are equal and symmetrical and therefore, can be deduced in terms of single state variable, $i_o(t)$. Similarly, the output capacitor voltage, $v_c(t)$ is the output voltage, $v_o(t)$. Based on the above discussion, $i_o(t)$ and $v_o(t)$ are taken as the state variables. The relationship between inductor currents and capacitor voltage with state variable are expressed in (2.55).

$$i_{Lf1}(t) = i_{Lf2}(t) = \frac{i_o(t)}{2}; \ v_c(t) = v_o(t) \tag{2.55}$$

In Fig. 2.25, the voltage across output filter inductor, L_{f1} is shown. Depending on the input voltage, $v_{hf}(t)$, the voltage across L_{f1} is divided into four part for one switching cycle (T_s). During each part, the state equations are derived which is subsequently averaged over complete switching cycle, T_s to obtain the final state equation. Assuming $L_{f1} = L_{f2} = L_o$, the state equations for different time duration are written as,

$$- v_o(t) = L_o \frac{di_{Lf1}(t)}{dt}; t_o \le t < t_1 \tag{2.56}$$

$$v_{ab}(t) - v_o(t) = L_o \frac{di_{Lf1}(t)}{dt}; t_1 \le t < t_2 \tag{2.57}$$

$$v_{ac}(t) - v_o(t) = L_o \frac{di_{Lf1}(t)}{dt}; t_2 \le t < t_3 \tag{2.58}$$

$$- v_o(t) = L_o \frac{di_{Lf1}(t)}{dt}; t_3 \le t < t_4 \tag{2.59}$$

It is to be noted that the durations $[t_o - t_1]$, $[t_1 - t_2]$, $[t_2 - t_3]$ and $[t_3 - t_4]$ exist for $\frac{t_o}{2}$, $\frac{t_\alpha}{2}$, $\frac{t_\beta}{2}$ and $\frac{T_s}{2}$ respectively. Averaging the Eqs. (2.56)–(2.59) over time period, T_s results in,

$$\underbrace{- v_o(t) \left(\frac{t_o}{2} + \frac{t_\alpha}{2} + \frac{t_\beta}{2} + \frac{T_s}{2} \right)}_{-v_o(t)T_s} + \underbrace{v_{ab}(t)t_\alpha + v_{ac}(t)t_\beta}_{\frac{3}{4}mV_mT_s} = L_o \frac{di_{Lf1}(t)}{dt} T_s \tag{2.60}$$

Further simplifying (2.60) results in

$$L_o \frac{di_o(t)}{dt} = \frac{3}{2} m(t) V_m - 2v_o(t) \tag{2.61}$$

where, $m(t)$ is the modulation index and V_m is the peak of the input phase voltage. The equivalent circuit for the output filter capacitor, C_o remains same for the switching cycle and therefore, the state-equation for capacitor voltage which is also the output voltage can be directly expressed as,

$$v_o(t) = \frac{1}{C_o} \int \left(i_o(t) - \frac{v_o(t)}{R_o} \right) \tag{2.62}$$

Differentiating (2.62) both sides results in,

$$C_o \frac{dv_o(t)}{dt} = i_o(t) - \frac{v_o(t)}{R_o} \tag{2.63}$$

To find the transfer functions, small perturbations in state variables, $v_o(t)$, $i_o(t)$ and $m(t)$ are introduced which are subsequently, put in (2.61) and (2.63). The *ac* and *dc* part of the both equations are equated. The Laplace transform of the *ac* part is carried out and thus, transfer functions are derived.

$$v_o(t) = V_o + \hat{v}_o; \ i_o(t) = I_o + \hat{i}_o; \ m(t) = M + \hat{m}; \tag{2.64}$$

Using (2.61), (2.63) and (2.64), the ac parts of the equations are equated and following equations are obtained.

$$L_o \frac{d\hat{i_o}}{dt} = \frac{3}{2}\hat{m}V_m - 2\hat{v_o} \tag{2.65}$$

$$C_o \frac{d\hat{v_o}}{dt} = \hat{i_o} - \frac{\hat{v_o}}{R_o} \tag{2.66}$$

Taking Laplace transform of (2.66), the V-I transfer function, $T_{p1}(s)$ is obtained.

$$T_{p1}(s) = \frac{v_o(s)}{i_o(s)} = \left(\frac{1/C_o}{s + \frac{1}{R_o C_o}} \right) \tag{2.67}$$

Further, taking Laplace transform of (2.65) and using (2.67), the $i_o - m$ transfer function, $T_{p2}(s)$ is obtained.

$$T_{p2}(s) = \frac{i_o(s)}{m(s)} = \frac{3V_m}{2L_o} \left(\frac{s + \frac{1}{R_o C_o}}{s^2 + \frac{s}{R_o C_o} + \frac{2}{L_o C_o}} \right) \tag{2.68}$$

2.7.2 Closed Loop Controller Design

In this subsection, control loop design of the converter is carried out for a given specification. The converter specification are given as follows: $V_m = 115\sqrt{2}$, $L_o = 1.2\,\text{mH}$, $C_o = 800\,\mu\text{F}$, $m = 0.73$, $R_o = 16.2\,\Omega$. The gain and phase plot of the transfer functions, $T_{p1}(s)$ and $T_{p2}(s)$ are shown in Fig. 2.26. The crossover frequency, f_{c2} of the transfer function, $T_{p2}(s)$ is 32 kHz and the phase margin is 90°. For the transfer function, $T_{p1}(s)$, the crossover frequency, f_{c1} is 200 Hz and the phase margin is found to be 93.5°. For loads such as battery and fuel cells, the current should be stable as their performance is very much dependent on current ripple. To have a smooth DC current in the output, it is necessary to have faster current control loop. Moreover, making the voltage loop significantly slower than current loop, design of controller become easier as both control loop can be designed independently. Figure 2.27 shows the control system for the proposed AC–DC converter. The voltage loop generates current reference. The bandwidth of current control loop is kept more than 10 times to have faster response compared to voltage control loop. This subsection is divided into three parts. In this first part, current control loop with appropriate PI controller is designed whereas in second part, the voltage control loop is designed. In the third part, the effect of load resistance on the controller performance is investigated.

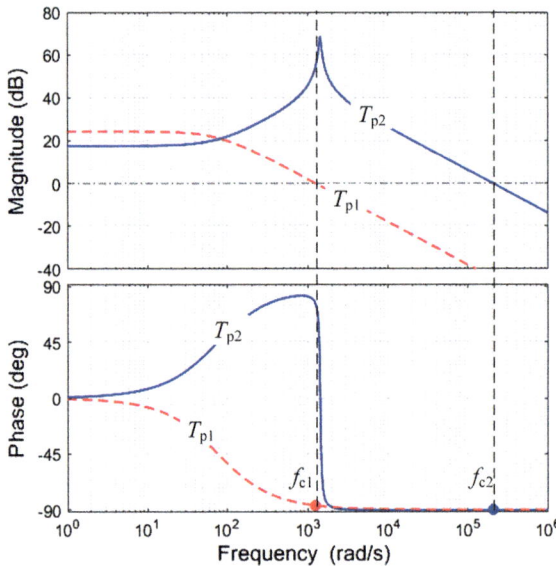

Fig. 2.26 Bode plot of the transfer functions, $T_{p1}(s)$ and $T_{p2}(s)$

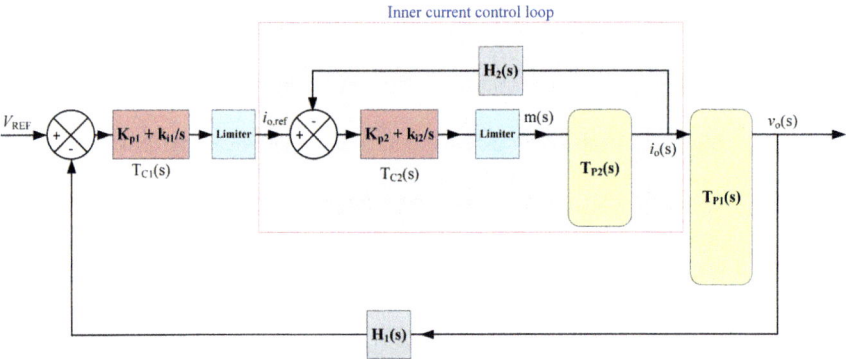

Fig. 2.27 l Bode plot of the transfer functions, $T_{p1}(s)$ and $T_{p2}(s)$

2.7.2.1   Current Control Loop

The objective of the design of current control loop is to meet the design criteria specified for bandwidth and phase margin. A PI controller is designed to increase the low frequency gain and reduce the steady state error between the reference inductor current and actual inductor current while maintaining a positive phase margin (typically 60°) and a large bandwidth.

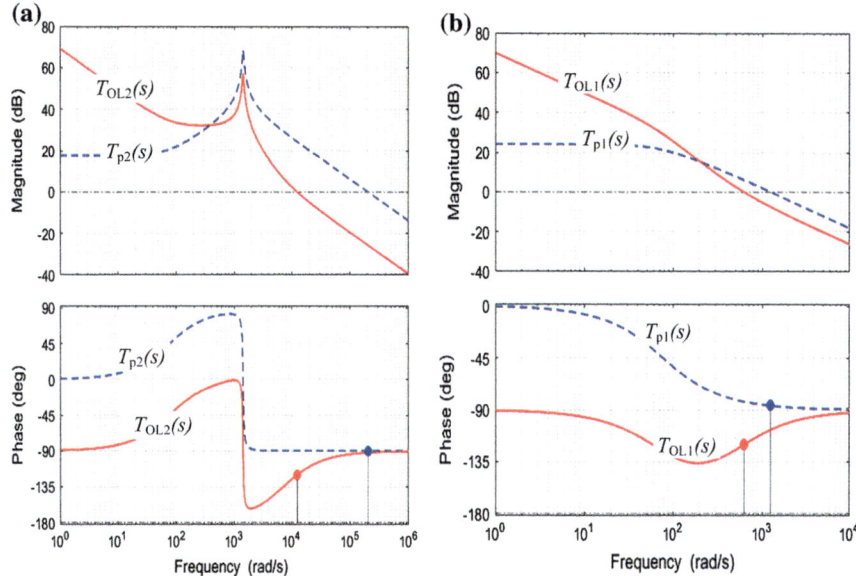

Fig. 2.28 **a** Bode plot of the transfer functions, $T_{p2}(s)$ and $T_{OL2}(s)$. **b** Bode plot of the transfer functions, $T_{p1}(s)$ and $T_{OL1}(s)$

Assuming the reference current to be 5.5 A, the gain of $H_2(s)$ block for 500 W output power and 90 V output voltage can be calculated. The gain of $H_2(s) = \frac{5.5}{(\frac{500}{90})} = 1$. From Fig. 2.27, the open loop transfer function of the current loop is given by,

$$T_{OL2}(s) = H_2(s)T_{c2}(s)T_{p2}(s) \tag{2.69}$$

For 60° phase margin at 2 kHz cross over frequency, the PI controller parameters are calculated based on gain and phase condition.

- The gain of $T_{OL2}(s)$ should be 1 at the crossover frequency i.e. $\mid T_{OL2}(s) \mid = 1$
- The phase of $T_{OL2}(s)$ at cross over frequency should be PM-180° i.e. $\angle T_{OL2}(s) = -120°$

Based on these two conditions, the PI controller parameters, k_{p2} and k_{i2} are calculated and found to be 0.053 and 383.34, respectively. The bode plot of $T_{OL2}(s)$ and $T_{p2}(s)$ are plotted in Fig. 2.28a. It is to be noted that the designed PI controller increases the low frequency gain. Moreover, the desired phase margin of 60° and 2 kHz cross over frequency is obtained.

2.7.2.2   Voltage Control Loop

The outer voltage loop is designed significantly slower than inner current control loop and therefore, the dynamics of current control loop are ignored while designing the voltage control loop. The outer voltage control loop generates the current reference for inner current control loop. For voltage control loop, the reference voltage is taken as 0.9. For 90 VDC output the gain, $H_1(s)$ is calculated to be $\frac{0.9}{90} = 0.01$. Thus, the open loop transfer function of the voltage control loop is given by,

$$T_{OL1}(s) = H_1(s)T_{c1}(s)T_{p1}(s) \tag{2.70}$$

For PI controller design, the phase margin of the $T_{OL1}(s)$ is considered to be 60° with cross-over frequency of 100 Hz. Similar to current control loop the PI parameters, k_{p1} and k_{i1} are designed based on the two conditions given by,

- The gain of $T_{OL1}(s)$ should be 1 at the crossover frequency i.e. $\mid T_{OL1}(s) \mid= 1$
- The phase of $T_{OL1}(s)$ at cross over frequency should be PM-180° i.e. $\angle T_{OL1}(s) = -120°$

Based on these two conditions, the PI controller parameters, k_{p1} and k_{i1} are calculated and found to be 40 and 19154, respectively. The bode plot of $T_{OL1}(s)$ and $T_{p1}(s)$ are plotted in Fig. 2.28b. It is to be noted that the designed PI controller increases the low frequency gain. Moreover, the desired phase margin of 60° and 2 kHz cross over frequency is obtained.

2.7.2.3   Effect of Output Load Variation on Control Performance

In this subsection, the open loop transfer function of both inner current control loop and outer voltage control loop is analysed for load variation. Figure 2.29 shows the bode plot for open loop transfer functions, $T_{OL1}(s)$ and $T_{OL2}(s)$ for load variation from 100–15%. The phase margin and cross over frequency remain unchanged for the load variation as shown in Fig. 2.29. Therefore, the designed control loops confirms the closed loop stability from full load to 15% of the full load.

2.7.3   Simulation Result

To validate the control loop design of the matrix based AC–DC converter, the converter is simulated in MATLAB-2016b with the specification shown in Table 2.3. The converter is designed for full load output power of 500 W ($R_o = 16.2\ \Omega$). A step load change from 15% of the full load to 100% of the full load is carried out to demonstrate the dynamic performance of the converter for the designed closed two loop control system.

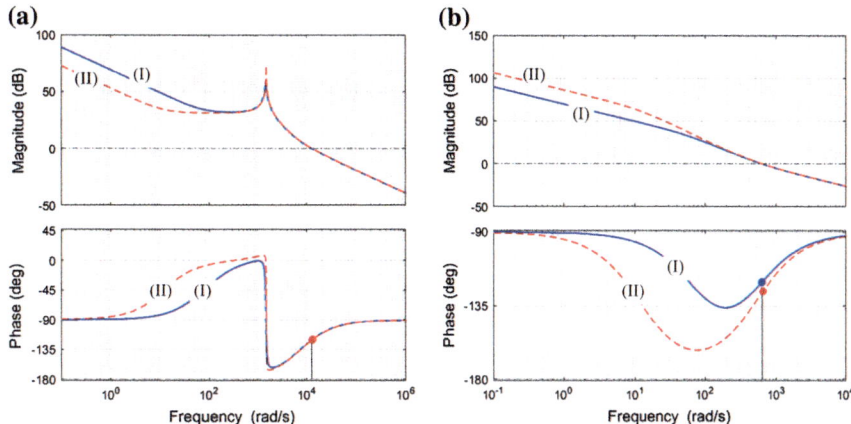

Fig. 2.29   a Bode plot of the transfer functions, $T_{OL2}(s)$ for full load (I) and 15% of the full load (II). **b** Bode plot of the transfer functions, $T_{OL1}(s)$ for full load (I) and 15% of the full load (II)

Fig. 2.30   a Output voltage, V_o in volts **b** Output current, I_o in amperes. **c** Phase-a input voltage, v_{an}. **d** Phase-a input current, i_a

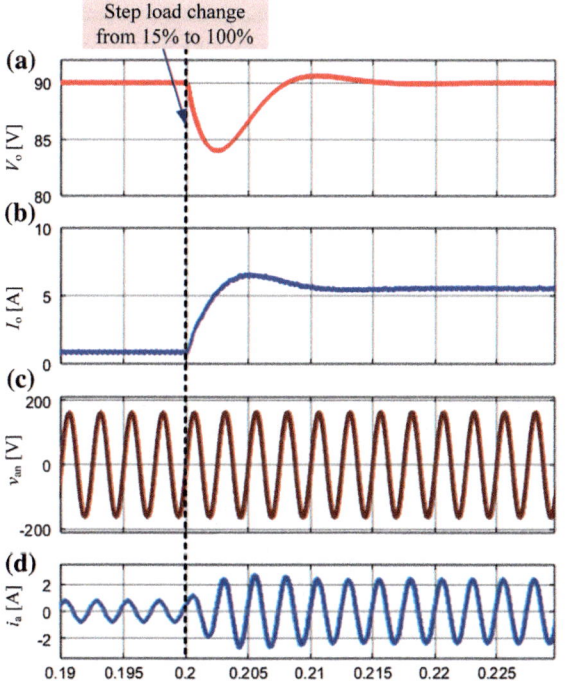

Figure 2.30 shows the simulated results of the proposed matrix based AC–DC converter for step load change under the designed closed loop control. At $t = 0.2$ s, the output load is suddenly increased. There is sudden undershoot in output voltage (6%) which gets regulated again at $t = 0.215$ s as shown in Fig. 2.30a. The corresponding variation in output current is shown in Fig. 2.30b. As the current control loop is faster, the overshoot is relatively smaller. The input phase voltage and phase current are shown in Fig. 2.30c, d.

2.8 Comparative Loss Evaluation of the Proposed Converter

In this section, the loss analysis of the proposed converter is carried out. This section is divided into three subsections. In the first subsection, the semiconductor losses of the proposed converter are carried out. In the second subsection, the semiconductor losses in the traditional six switch buck rectifier based on [19] are carried out. Subsequently, the semiconductor losses of these two converters are compared in the third subsection at different switching frequency and at different output power. It is to be noted that since the input and output filter of the two converters are almost identical, the loss comparison is carried out based on only total semiconductor losses.

2.8.1 Semiconductor Losses in the Proposed Converter

The semiconductor losses in the proposed converter are divided into two parts- (1) conduction loss and (2) switching loss. These two losses are calculated for the proposed converter as follows:

2.8.1.1 Conduction Loss

In the proposed modulation scheme, the body diodes of the MOSFETs in the matrix switches do not conduct and therefore, the switch conduction losses is only due to the ON resistance of the MOSFETs. The switch rms current is given by (2.19). The total conduction loss in the matrix switches, $P_{c,Matrix}$ is given by,

$$P_{c,Matrix} = 12 i_{SW,rms}^2 R_{ds,ON} \tag{2.71}$$

where, $R_{ds,ON}$ is the ON resistance of the MOSFETs used to implement matrix switch.

Similarly, the conduction loss in the CDR diodes is calculated based on *average* and *rms* current calculated in (2.22) and (2.23). The total conduction loss in the CDR diodes, $P_{c,CDR}$ is given by,

$$P_{c,CDR} = 2(i_{D,rms}^2 R_D + i_{D,avg} V_f) \tag{2.72}$$

where, R_D and V_f are the ON resistance and forward voltage drop of the CDR diodes, respectively.

2.8.1.2 Switching Loss

The switching losses in the proposed converter are divided into three parts. The first part is due to the overlapping of voltage and current during MOSFET turn ON and turn OFF. The second part is due to the charging and discharging of the parasitic capacitance of the MOSFETs and the CDR diodes. The third part is the gate driving loss. To calculate the total switching loss of the proposed converter, all these three losses are taken into consideration. As SiC diodes are used in both proposed and traditional, the reverse recovery losses have not been considered for switching loss calculation in diodes.

Taking $V_{DS,ON}$ as drain-source voltage prior to turn ON of the MOSFET and I_{ON} as the drain-source current after turn ON, the power loss during turn ON, $P_{sw,ON}$ can be approximated by the following:

$$P_{sw,ON} = \frac{V_{DS,ON} I_{ON}}{2} t_r f_s \tag{2.73}$$

where, t_r is the overlapping period during the turn ON of the MOSFET of the matrix switch.

The drain-source voltage, $V_{DS,ON}$ varies throughout the line period, $T_i = \frac{1}{f_i}$ and thus, the average value of $V_{DS,ON}$ is taken for calculating turn ON loss using (2.75). The current, I_{ON} remains constant and its value is given by $\frac{I_o}{2}$, where I_o is the output current of the converter.

Figure 2.31 shows the voltage variation, $v_{sw}(\theta)$ across one of the MOSFETs of the matrix switch. During the one third of the line period, ($\theta = \frac{4\pi}{3}$ to 2π) the voltage across the MOSFET is zero which results in ZVS in this region. The average voltage across the MOSFET is thus calculated by integrating the switch voltage, v_{sw} from $\theta = 0$ to $\theta = 2\pi$.

If the average value of the drain-source voltage is represented by $< V_{DS,ON} >$, then

$$< V_{DS,ON} > = \frac{1}{2\pi} \int_o^{2\pi} v_{sw}(\theta) d\theta = \frac{3\sqrt{3}}{\pi} V_m \tag{2.74}$$

The total turn ON power loss, $P_{sw,ON,total}$ for the matrix switches is given by the following:

$$P_{sw,ON,total} = \frac{9\sqrt{3}}{4\pi} V_m I_o t_r f_s \tag{2.75}$$

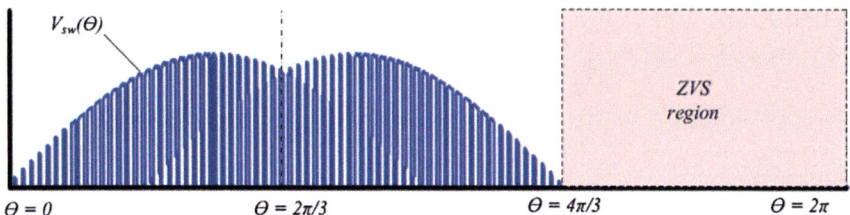

Fig. 2.31 Voltage across on of the MOSFETs of the matrix switch for a line cycle ($T_i = \frac{1}{f_i}$)

During the turn OFF the matrix switch, the voltage across the two back-to-back connected MOSFETs are different. One of the MOSFETs is charged to $V_{DS,ON}$ whereas other MOSFET is charged to the forward voltage drop of the diode, V_f. As V_f is too small in comparison to $V_{DS,ON}$, the turn OFF loss due to one MOSFET is only taken into consideration. With this, the total turn OFF loss of the matrix switch is given by,

$$P_{sw,OFF,total} = \frac{9\sqrt{3}}{4\pi} V_m I_o t_f f_s \tag{2.76}$$

where, t_f is the overlapping period during the turn OFF of the MOSFET.

The switching loss due to charging and discharging of the output capacitor in each of the MOSFETs of the matrix switch is derived as,

$$P_{sw,Matrix} = \frac{C_{oss} f_s}{2\pi} \int_0^{2\pi} v_{sw}^2(\theta) d\theta \tag{2.77}$$

where, C_{oss} is the output capacitance of the MOSFET of the matrix switch. The output capacitance of the diodes of the CDR circuit, $C_{oss,D}$ also charges and discharges resulting in additional switching loss. The voltage across one of the CDR diodes, D_1 is shown for a switching cycle, T_s in Fig. 2.32. The diode capacitance charges and discharges during this period which results in switching loss. The switching loss at given θ during a sector depends on the voltages v_{ab} and v_{ac}. Integrating the loss at theta for a full sector provides loss incurred by the CDR diode, D_1,

$$P_{sw,CDR}(\theta) = C_{oss,D}(v_{ab}^2 + v_{ac}^2) f_s \tag{2.78}$$

Integrating the (2.78) for $\theta = \frac{\pi}{3}$ to $\frac{2\pi}{3}$ results in,

$$P_{sw,CDR} = \frac{3}{\pi} \int_{\frac{\pi}{3}}^{\frac{2\pi}{3}} P_{sw,CDR}(\theta) d\theta = \frac{3}{\pi} \int_{\frac{\pi}{3}}^{\frac{2\pi}{3}} C_{oss,D}(v_{ab}^2 + v_{ac}^2) f_s d\theta \tag{2.79}$$

which is simplified to,

Fig. 2.32 Voltage across on of the CDR diodes, D_1 for a switching cycle, T_s

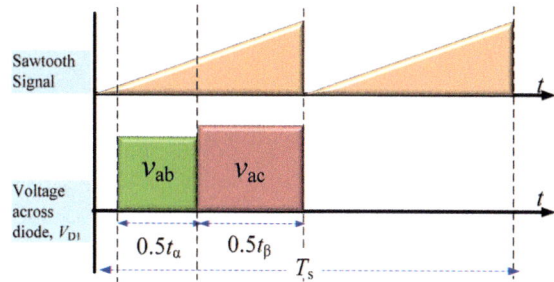

$$P_{sw,CDR} = C_{oss,D} V_m^2 \left(3 + \frac{9\sqrt{3}}{4\pi} \right) f_s \qquad (2.80)$$

where, $C_{oss,D}$ is the output capacitance of the diode of the CDR circuit. As there are twelve MOSFETs and two CDR diodes in the proposed converter, the total switching loss due to charging and discharging of the MOSFETs capacitance and the CDR diodes is given by,

$$P_{sw,C,total} = \frac{12 \left(\pi + \frac{3\sqrt{3}}{8} \right)}{\pi} C_{oss} V_m^2 f_s + 2C_{oss,D} V_m^2 \left(\frac{1}{6} + \frac{\sqrt{3}}{8\pi} \right) f_s \qquad (2.81)$$

where, C_{oss} is the output capacitance of each of the MOSFETs. Similarly, the total gate drive loss of the proposed converter is given by,

$$P_{sw,G,total} = 12 C_{iss} V_g^2 f_s \qquad (2.82)$$

where, C_{iss} is the input capacitance of each of the MOSFETs.

2.8.1.3 Total Loss

Taking all losses into consideration, the total semiconductor loss, $P_{total,proposed}$ of the matrix switcher under the proposed modulation scheme is given by the following:

$$P_{total,proposed} = \underbrace{\frac{3m I_o^2 R_{ds,ON}}{\pi} + I_o^2 R_D + I_o V_f}_{conduction\ loss} + P_{switchingloss,proposed} \qquad (2.83)$$

where,

$$P_{switchingloss,proposed} = \frac{9\sqrt{3}}{4\pi} V_m I_o f_s \left(t_{tr} + t_{tf}\right) + \frac{12\left(\pi + \frac{3\sqrt{3}}{8}\right)}{\pi} C_{oss} V_m^2 f_s +$$

$$12 C_{iss} V_g^2 f_s + 2 C_{oss,D} V_m^2 \left(\frac{1}{6} + \frac{\sqrt{3}}{8\pi}\right) f_s$$

$$(2.84)$$

2.8.2   Semiconductor Losses in the Traditional Six Switch Buck Rectifier

The comprehensive loss analysis of the traditional six switch buck rectifier is presented in [19]. The *rms* and *avg* current of the semiconductor devices are derived based on [19]. It should be noted that in the traditional buck rectifier, the amplitude of the switch current is equal to the load current, I_o unlike the proposed converter where the amplitude of the switch current is half of the load current. This not only reduces the switch *rms* current but also reduces the switching loss almost by half in the proposed converter. Assuming all the semiconductor devices of the traditional converter are identical to the proposed converter, the total loss of the traditional six switch buck rectifier is given by the following:

$$P_{total,traditional} = \underbrace{\frac{6}{\pi} M R_{ds,ON} I_o^2}_{\text{Mosfet conduction loss}} + \underbrace{\frac{6}{\pi} M (I_o^2 R_D + I_o V_f)}_{\text{Switch diode conduction loss}} +$$

$$\underbrace{\left(1 - \frac{3M}{\pi}\right) (I_o^2 R_f + I_o V_f)}_{\text{Output diode conduction loss}} + P_{switchingloss,traditional} \qquad (2.85)$$

$$P_{switchingloss,traditional} = \frac{3\sqrt{3}}{\pi} V_m I_o t_r f_{st} +$$

$$\frac{12\left(\frac{7\pi}{16} - \frac{9\sqrt{3}}{32}\right)}{\pi} \left(\frac{C_{oss} + 3C_{oss,D}}{2}\right) V_m^2 f_{st} + 6 C_{iss} V_g^2 f_{st} \qquad (2.86)$$

where, M and f_{st} are the modulation index and the switching frequency of the traditional converter, respectively. It is worth noticing that all the diodes of the traditional converter are assumed to be identical.

It is important to note here that the value of modulation index is not same in both the proposed and traditional converter. For same input and output specification, the modulation indexes and the switching frequencies of the proposed and traditional converter are related by,

$$m = 2M; \; f_{st} = 2f_s \tag{2.87}$$

The proposed converter generates bipolar symmetrical high frequency AC of 40 kHz frequency which is equivalent to 80 kHz output, if rectified. As buck converter generates unipolar high frequency output at its output, the matrix switching frequency f_s and the buck rectifier switching frequency, f_{st} are related as shown in (2.87).

2.8.3 Snubber Loss in the Proposed Converter and Six Switch Buck Rectifier

The snubber loss equation for both topologies have been derived. For the 12 MOSFETs in the proposed converter, 12 R-C snubber circuits are added whereas in the six switch buck rectifier 6 MOSFETs requires 6 R-C snubber circuit. It has to be noted that for the same input line frequency ripple, the buck rectifier switches has to be operated at twice the switching frequency as compared to matrix based buck rectifier. With this, the snubber loss equation can be derived as,

$$P_{snubber,proposed} = 12C_s V_m^2 f_s \left(1 + \frac{3\sqrt{3}}{8\pi}\right) \tag{2.88}$$

$$P_{snubber,buck} = 6C_s V_m^2 f_s \left(1 + \frac{3\sqrt{3}}{4\pi}\right) \tag{2.89}$$

2.8.4 Comparison of the Semiconductor Losses in the Proposed and the Traditional Six Switch Buck Rectifier

In this subsection the semiconductor losses in the proposed and the traditional six switch converter are compared at different output power and switching frequencies. For comparing the losses of the proposed and the traditional converter, identical semiconductor devices have been chosen. Based on the selected semiconductor devices given in Table 2.6, the various parameters of the two converters are shown in Table 2.5.

Assuming the device parameters to be same for 100−2000 W, the total semiconductor losses including snubber losses of the two converters are calculated for varying switching frequency, f_s (20−100 kHz) evaluated using the loss equations derived in Sect. 2.7.1 for modulation index, $m = 0.7$. The results are illustrated through a 3D plot in Fig. 2.33. The followings observation are made through the loss analysis of the proposed and the traditional buck converter [19]:

Table 2.5 Semiconductor device parameters

Parameter	Value
Mosfet ON resistance, $R_{ds,ON}$	200 mΩ
Mosfet input capacitance, C_{iss}	2170 pF
Mosfet output capacitance, $C_{oss,D}$	70 pF
CDR diode Resistance, R_D	50 mΩ
CDR diode capacitance, $C_{oss,D}$	200 pF
Forward voltage drop of CDR diode, V_f	1.8 V
Snubber capacitor, C_s	0.5 nF

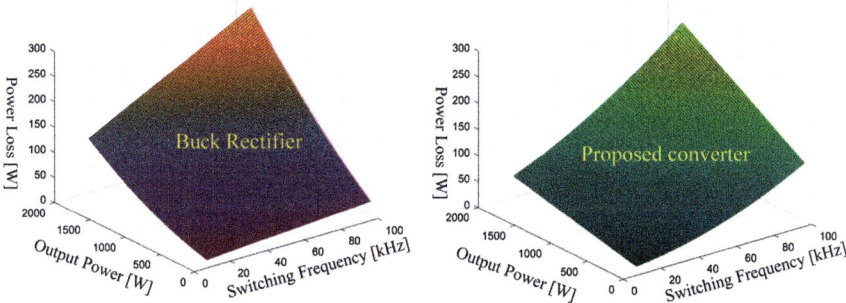

Fig. 2.33 Total semiconductor power loss versus output power, P_o and switching frequency, f_s for the proposed rectifier and six switch buck rectifier [19]

- The proposed converter provides lower semiconductor loss compared to the three phase buck rectifier for the same input-output specifications at relatively lower switching frequency and higher output power;

- For a given output power, the semiconductor losses of both - the traditional and the proposed converter becomes equal at certain switching frequency. For example, at $f_s = 40$ kHz, the losses of these two converters are almost equal at 500 W output power. However, increasing the output power at this switching frequency results in lower semiconductor losses in the proposed converter than the three phase buck rectifier as shown in Fig. 2.33; and

- As shown in Fig. 2.33, by increasing the output power, the yellow line shifts at higher switching frequencies which shows that with the increase in the output power, the proposed converter can be operated at higher switching frequency with overall lower semiconductor loss than the traditional three phase six switch buck rectifier.

- The proposed converter can be operated at higher switching frequency with lower semiconductor loss at higher power as shown in Fig. 2.33. As the increased switching frequency corresponds to reduced size/volume of the passive and magnetic elements, the proposed converter can provide higher *power density* at high output power than the traditional six switch buck rectifier.

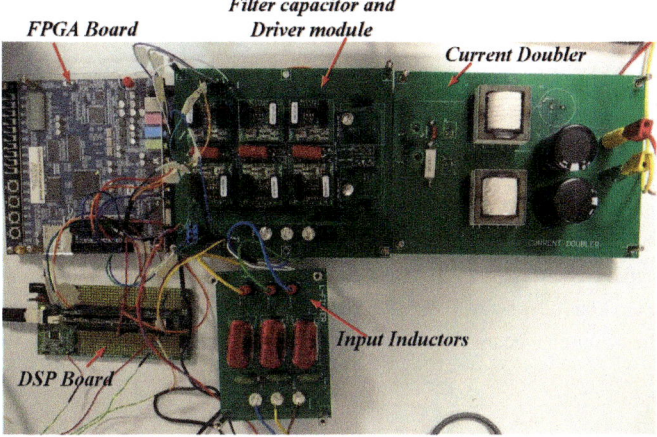

Fig. 2.34 Hardware prototype of the proposed matrix based AC–DC converter

2.9 Experimental Verification

2.9.1 Prototype and Structure

Figure 2.34 shows the prototype of the proposed matrix based non isolated AC–DC rectifier. The converter is divided into four modules. First module consists of filter inductor and damping resistor. The second module consists of power devices, gate drivers and input filter capacitor. The output of second module is processed using current doubler (3rd module) to provide the rectified DC output voltage. The fourth module consists of voltage/ current sensors, DSP board and FPGA board. The matrix switches are formed using two discreet MOSFETs by connecting them back to back as shown in Fig. 2.4. The three phase input voltage is sensed and given to DSP board to generate PWM signals which are further processed using FPGA board to generate switching signal for gate drivers [21]. The active and passive components selected for the experimental validation of the proposed converter is given in Table 2.6.

2.9.2 Controller

The controller hardware for the proposed converter includes a DSP board and a FPGA board. The DSP board used is the TI TMDSCNF28335 evaluation board whereas the FPGA board used is ALTERA QUARTUS II board. The three phase input AC voltage is sensed and given to the DSP board which in turn, generates PWM signal. A abc-dq based PLL is implemented inside DSP for calculating the phase angle, θ. Based on θ and modulation index, m the DSP generates PWM signals. The PWM

Table 2.6 Active and passive components selected for experimental validation of the proposed matrix based three phase AC–DC converter

	Component	Specification
1	MOSFET, S_1-S_6	FCA16N60N, 600 V, 16 A
2	Diode D_1, D_2	C3D10060A-ND, 600 V 14.5 A
3	Input filter inductors, L_a, L_b, L_c	513–1660-ND, 200 μH 7 A
4	Input filter capacitor, C_a, C_b, C_c	PCF1569-N, 1.2 μF 630 VDC
5	Output inductor, L_{f1}, L_{F2}	513–1654-ND, 1.2 mH 4.7 A
6	Output capacitor, C_o	Electorlytic capacitor 400 μF, 400 V
7	FPGA controller board	ALTERA QUARTUS II
8	Microcontroller board	TMDSCN28335

signals are further processed using FPGA board based on the sector information and thus, final switching signals are generated which are given to the gate drivers [22]. The detailed digital control implementation is discussed in the next chapter.

2.9.3 Experimental Results

2.9.3.1 Key Experimental Results

The proposed matrix based AC–DC converter is tested with a resistor load and three phase input voltage of 115 V(rms), AC at 400 Hz. The power devices are mounted with heat sinks and additional fan cooling is provided for the converter to test at full load. Figure 2.35 shows input phase-a voltage, v_{an} and high frequency AC, v_{hf}. The proposed matrix topology converts three phase line frequency AC directly into single phase high frequency AC. Subsequently, the high frequency AC output is processed using current doubler rectifier to provide the required DC output voltage. Figure 2.36 shows the rectified output DC voltage, V_o and high frequency AC output, v_{hf} of the matrix converter.

Figure 2.37 shows the high frequency AC output the converter. It is evident that the output is symmetrical and bipolar. The small zero period between the adjacent voltage is due to the presence of dead time, t_d which is required to avoid the short circuit of the input filter capacitor during switch transition. Figure 2.38 shows the high frequency AC output, v_{hf} and high frequency AC current, i_{hf} at full load. A small R-C snubber (50 Ω and 1.5 nF) is kept across the input of current doubler to reduce the voltage spikes. Because of parasitic capacitance of current doubler diodes, the high frequency AC is having some spikes which has further scope of improvement by the proper design of PCB layout. Figure 2.39 shows the input phase voltage, V_{an} and the voltage across switch SW_{11}. The switch voltage rises from zero become maximum and then falls to zero. For a 60° time interval, the voltage

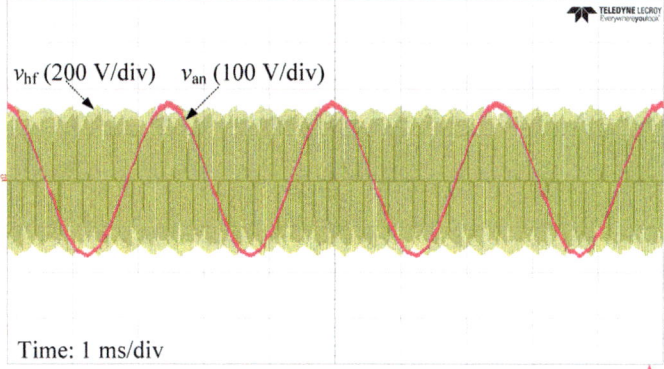

Fig. 2.35 Input phase voltage, v_{an} and high frequency AC output, v_{hf}

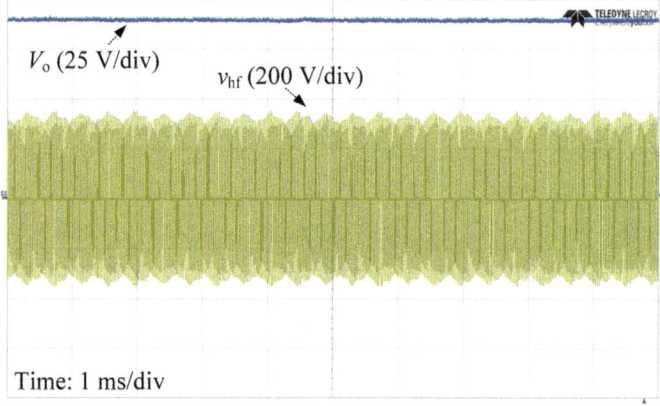

Fig. 2.36 Output DC voltage, V_o and High frequency AC output, v_{hf}

across the switch remains zero. Figure 2.40 shows the voltage across switch SW_{11} and SW_{12}. Both switches SW_{11} and SW_{12} forms S_1 as shown in Fig. 2.4. Each of the two switches are ON For 60° time interval. It is because of the forward biasing of the switch diode. Consequently, it gives rises to zero switching loss during this particular interval and therefore, improves power conversion efficiency. Therefore, even though the proposed matrix converter has 12 switches, half of the switches will not contribute to the switching loss because of the zero voltage across them. Figure 2.41 shows the phase-a voltage, v_{an} and phase current, i_a at full load. The displacement power factor of the converter is close to unity. Figure 2.42 shows the three phase input current. The currents are balanced and symmetrical. The THD of the current is estimated for full load and 50% of the full load. It is found to be 3.7 and 4.19%. Figure 2.43 shows the voltage across current doubler diodes D_1 and D_2. Each of the diodes remains ON for less than half of the switching cycle. During the

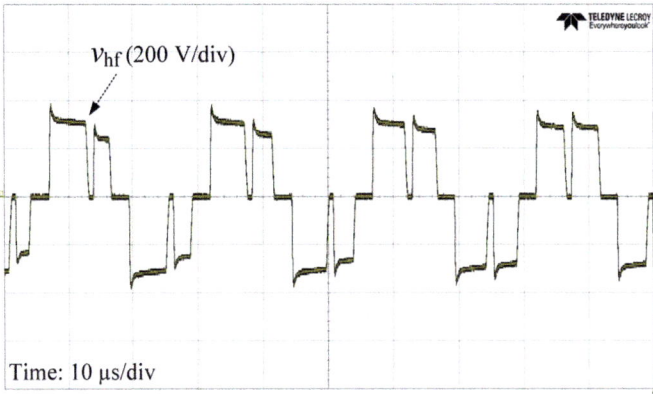

Fig. 2.37 High frequency AC output, v_{hf} of the matrix converter

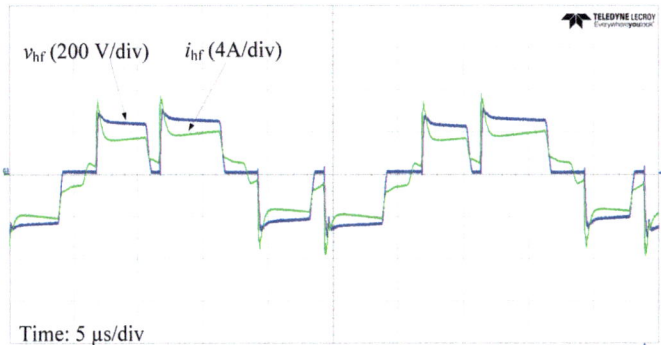

Fig. 2.38 High frequency AC output, v_{hf} and high frequency AC current, i_{hf} of the matrix converter at full load

mode when power is not delivered from source to load, both of the diodes D_1 and D_2 conducts and zero voltage appears across the input of current doubler circuit.

2.9.3.2 Discussion on the Power Quality of the Proposed Converter

The proposed power converter provides superior input power quality. Figure 2.44 shows the THD variation of input current at different load. Even at 20% of the full load, THD is better that 7%. The THD at full load is found to be 3.7%. The THD of the converter can be further improved by increasing the input filter capacitor. However, the maximum value of input filter capacitor is limited by displacement power factor. The proposed converter is tested at 500 W power with 3.7% of THD and almost unity power factor that demonstrates the performance of the converter in terms of the input power quality. Figure 2.45 shows the variation of displacement power factor with the

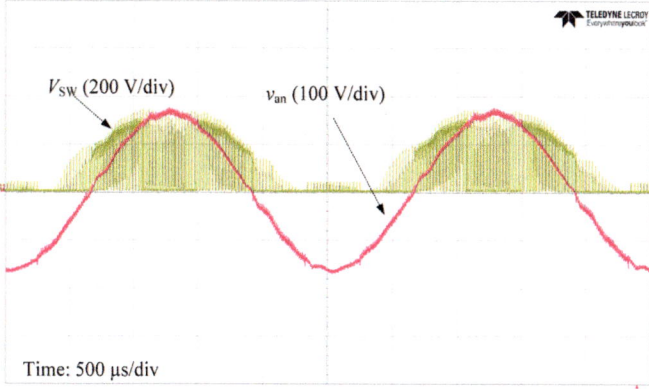

Fig. 2.39 Input phase-a voltage, v_{an} and switch SW_{11} voltage

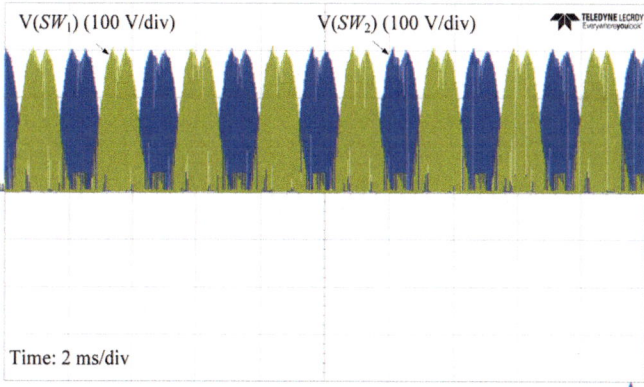

Fig. 2.40 Voltage across switch SW_{11} and SW_{12}

output load. The displacement power factor is almost unity at 100% of the load. It reduces with reduction in the output load. However, even at 50% of the full load, the displacement power factor is more than 90%.

2.9.3.3 Discussion on the Power Conversion Efficiency of the Proposed Converter

The proposed converter is tested with resistive load and programmable three phase power supply. As shown in the modes of operation, there is no switch diode conduction which contributes in reducing conduction losses of the switches. Moreover, the use of current doubler rectifier circuit reduces switch rms current by half which further reduces the switch conduction loss. To realize the matrix topology, each matrix switch is realized by two back to back connected switches. During the switching

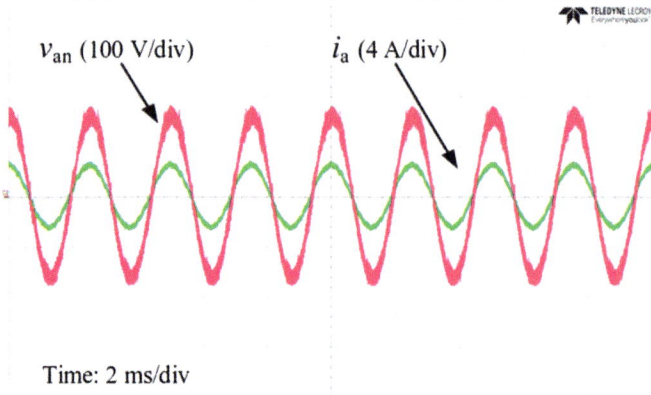

Fig. 2.41 Input phase-*a* voltage, v_{an} and phase current, i_a. The displacement power factor is close to unity

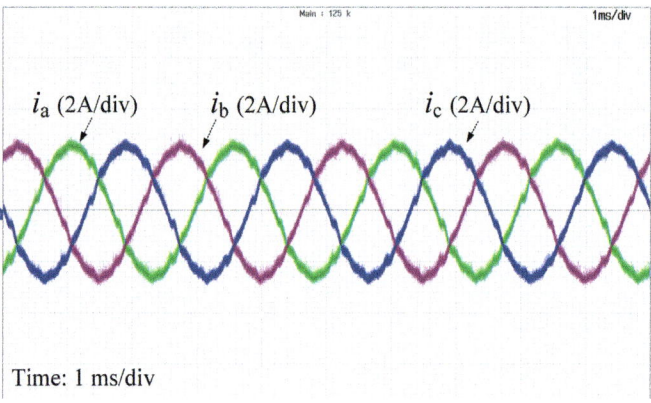

Fig. 2.42 Three phase input current, i_a, i_b, i_c. The Total Harmonic Distortion (THD) of the current is found to be 3.7%

process, one of the two switches undergoes natural ZVS as it remain forward biased which contributes in reducing the switching loss. The total semiconductor losses have been calculated in Sect. 2.8 for the proposed converter. The input and output filter losses are calculated according to their data-sheet. The effect of ripple voltage and ripple current is ignored in the total loss calculation of the proposed converter. A R-C snubber of $50\,\Omega$ and $0.5\,\text{nF}$ has been added across all the switches of the matrix converter to reduce the high frequency voltage ringing due to the parasitic effects. The theoretical loss distribution of the proposed converter is shown in Fig. 2.46. The total estimated loss of the converter at $500\,\text{kW}$ output power is $43.09\,\text{W}$ and therefore, the theoretical full load efficiency of the proposed converter is found to be 92.1%.

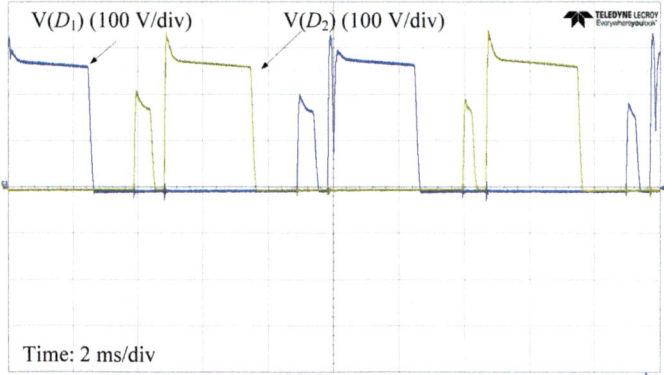

Fig. 2.43 Voltage across the diodes, D_1 and D_2

Fig. 2.44 Experimental
THD at different output load

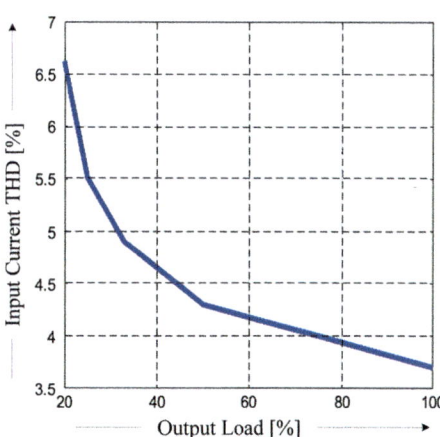

Fig. 2.45 Displacement
power factor (DPF) at
different output load

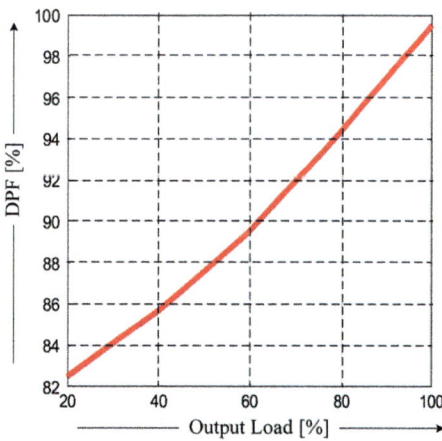

Fig. 2.46 Theoretical
distribution of the power loss
in watts for the proposed
converter. 1: Conduction loss
2: Switching loss 3: Input
filter loss 4: Output filter loss
5: Snubber loss

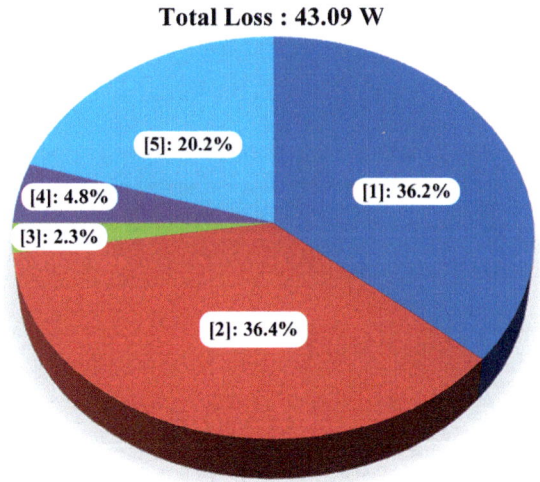

The experimental full load efficiency of the prototype was found be 91% which is very close to the theoretical efficiency of the converter. The difference in theoretical and experimental efficiency is due to the parasitic elements and voltage and current ripple which have been ignored in the calculation. The operation of proposed converter at high switching frequency ($\geq 40\,$kHz) requires proper layout design for reducing the parasitic effects. There is further scope of optimizing the power conversion efficiency by selecting semiconductor devices and PCB layout design. However, the main focus of this chapter is the topology, modulation techniques and its experimental validation; the optimization of efficiency is not carried out.

2.10   Benchmarking of the Proposed Converter

The quantifications of the power density, efficiency, power quality have been carried out for all the three types of power converters and tabulated in Table 2.7. The values shown in the table are calculated for six switch buck rectifier and the proposed matrix based buck rectifier for the same input and output specifications. The ATRU specifications are chosen from the product datasheet given in [23]. The components chosen in both the converters are identical. For ATRU, the values have been obtained from the product datasheet. It is evident from the Table 2.7 and Fig. 2.47 that the proposed matrix based AC–DC converter provides more power density as well as is more efficient than both ATRU and six switch buck rectifier.

It is to be noted that the proposed converter produces input current ripple at twice the switching frequency of the matrix switches. In a six switch buck rectifier, the input current ripple frequency is same as the switching frequency. Therefore, for the same ripple magnitude, the size of the inductor in six switch buck rectifier is two

Table 2.7 Benchmarking for nonisolated matrix based buck rectifier

		Type	Power density(W/Kg)	THD(%)	Efficiency(%)
1	Proposed converter	Active	**526.34**	**3.7**	**94.17**
2	ATRU	Passive	468.75	<5	93
3	Six switch buck rectifier	Active	476.59	<5	93.87

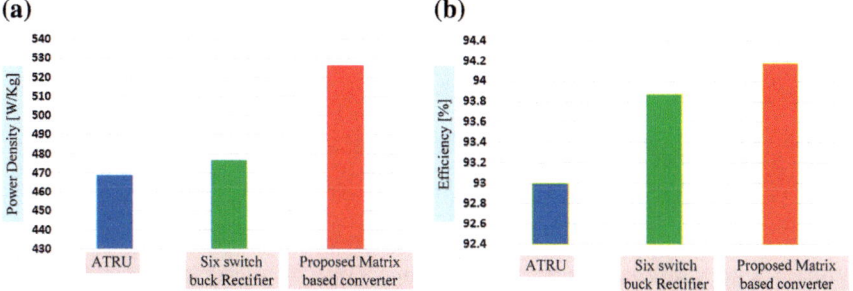

Fig. 2.47 Benchmarking for the nonisolated matrix based AC–DC converter. **a** Power density [W/kg] **b** Efficiency [%]

times of the proposed matrix based buck rectifier and thus, the proposed rectifier is better in terms of power density than the six switch buck rectifier. Further, the ATRU requires line frequency multi-pulse auto transformer which contributes it to have the lowest power density in all the three.

The proposed converter shows highest power conversion efficiency. Even though, the proposed converter has twice the number of MOSFETs compared to six switch buck rectifier, the Zero Voltage Switching in half of the MOSFETs and no body diode conduction results in lower switch loss. Moreover, the MOSFETs in the proposed converter switches with half of the current as compared to a six switch buck rectifier. Thus the overall power losses in the proposed converter is less leading to improved efficiency. ATRU has line frequency transformer and required passive diode for rectification. It shows the lowest efficiency. In Fig. 2.48, the efficiency of the proposed converter and six switch buck rectifier is shown for 100, 50 and 25% of the output load. The proposed converter provides better power conversion efficiency than the six switch buck rectifier for 100–25% of the output load. However, the difference in efficiency reduces with reduced output load.

Due to simple structure and no switching devices, the ATRU shows highest reliability in all the three. However, with new state of the art semiconductor devices such as SiC and GaN, the active rectifiers reliability can be improved at par with the ATRU in the near future.

Fig. 2.48 Efficiency of the proposed converter and six switch buck rectifier at 100, 50 and 25% of the output load

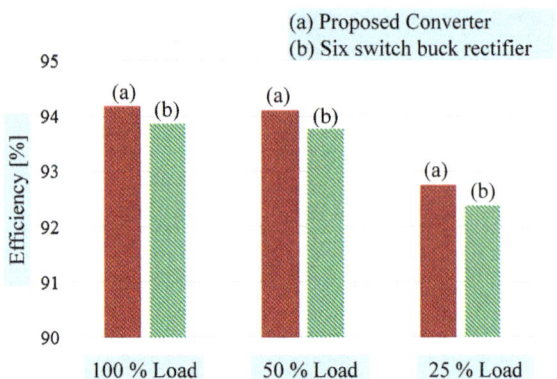

2.11 Conclusion

In this chapter, a novel matrix based non-isolated three phase AC–DC converter suitable for aircraft system has been presented. The matrix converter allows the use of CDR circuit and thus, steps down the voltage gain by two for the same modulation index. For lower output voltage, a conventional three phase buck rectifier must be operated at lower modulation index which in turn increases switch rms current and reduces power conversion efficiency. The proposed topology is very suitable for the applications where lower voltage gain $\left[\left(\frac{V_o}{V_{in,rms}}\right)\right]$ is required. The proposed SVM based modulation scheme is digitally implemented using DSP and FPGA for generating intermediated 40 kHz high frequency AC output. Further, the input current THD is demonstrated below 5% through hardware experiments at full load. Even at 20% of the load, the current THD only degrades to 6.5%. As the converter is operating at high switching frequency and discrete semiconductor switches are used to realize matrix switches, the PCB layout design is very important for reducing the parasitic effects. The proposed converter provides superior input power quality (THD = 3.7%), unity DPF and power conversion efficiency of 91% at 500 W output power. For 20% of the load, the DPF is shown experimentally to be 82%. With small input filter and output filter, the proposed topology has potential to provide high power density. The use of current doubler in the output side reduces the output current ripple by interleaving the output inductors. Comprehensive simulation followed by hardware experiments on hardware prototype validates the feasibility and suitability of the proposed converter for aircraft systems.

References

1. T. Nussbaumer, M. Baumann, J. Kolar, Comprehensive design of a three-phase three-switch buck-type pwm rectifier. IEEE Trans. Power Electron. **22**, 551–562 (2007)
2. J. Kolar, T. Friedli, The essence of three-phase pfc rectifier systems-part i. IEEE Trans. Power Electron. **28**, 176–198 (2013)

3. T. Friedli, M. Hartmann, J. Kolar, The essence of three-phase pfc rectifier systems-part ii. IEEE Trans. Power Electron. **29**, 543–560 (2014)
4. S. Ratanapanachote, H.J. Cha, P. Enjeti, A digitally controlled switch mode power supply based on matrix converter, in *2004 IEEE 35th Annual Power Electronics Specialists Conference, 2004. PESC 04*, vol. 3 (2004), pp. 2237–2243
5. J. Conde-Enriquez, J. Benitez-Read, J. Duran-Gomez, J. Pacheco-Sotelo, Three-phase six-pulse buck rectifier with high quality input waveforms. IEE Proc. Electr. Power Appl. **146**, 637–645 (1999)
6. J. Doval-Gandoy, C. Penalver, Dynamic and steady state analysis of a three phase buck rectifier. IEEE Trans. Power Electron. **15**, 953–959 (2000)
7. S.-B. Han, N.-S. Choi, C.-T. Rim, G.-H. Cho, Modeling and analysis of static and dynamic characteristics for buck-type three-phase pwm rectifier by circuit dq transformation. IEEE Trans. Power Electron. **13**, 323–336 (1998)
8. Y. Nishida, T. Kondoh, M. Ishikawa, K. Yasui, Three-phase pwm-current-source type pfc rectifier (theory and practical evaluation of 12kw real product), in *Proceedings of the Power Conversion Conference, 2002. PCC-Osaka 2002*, vol. 3 (2002), pp. 1217–1222
9. Y. Sato, T. Kataoka, State feedback control of current-type pwm ac-to-dc converters. IEEE Trans. Ind. Appl. **29**, 1090–1097 (1993)
10. M. Su, H. Wang, Y. Sun, J. Yang, W. Xiong, Y. Liu, Ac/dc matrix converter with an optimized modulation strategy for v2g applications. IEEE Trans. Power Electron. **28**, 5736–5745 (2013)
11. B. Guo, F. Wang, R. Burgos, E. Aeloiza, Control of three-phase buck-type rectifier in discontinuous current mode, in *2013 IEEE Energy Conversion Congress and Exposition (ECCE)* (2013), pp. 4864–4871
12. E. Sanchis, E. Maset, J. Carrasco, J. Ejea, A. Ferreres, E. Dede, V. Esteve, J. Jordan, R. Garcia-Gil, Zero-current-switched three-phase svm-controlled buck rectifier. IEEE Trans. Ind. Electron. **52**, 679–688 (2005)
13. S. Bassan, G. Moschopoulos, Zero-current-switching techniques for buck-type ac-dc converters, in *29th International Telecommunications Energy Conference, 2007. INTELEC 2007* (2007), pp. 506–513
14. H. Krishnamoorthy, P. Garg, P. Enjeti, A matrix converter-based topology for high power electric vehicle battery charging and v2g application, in *IECON 2012 - 38th Annual Conference on IEEE Industrial Electronics Society* (2012), pp. 2866–2871
15. V. Vlatkovic, D. Borojevic, F.C. Lee, A zero-voltage switched, three-phase isolated pwm buck rectifier. IEEE Trans. Power Electron. **10**, 148–157 (1995)
16. R. Jain, N. Mohan, R. Ayyanar, R. Button, A comprehensive analysis of hybrid phase-modulated converter with current-doubler rectifier and comparison with its center-tapped counterpart. IEEE Trans. Ind. Electron. **53**, 1870–1880 (2006)
17. P. Alou, J.A. Oliver, O. Garcia, R. Prieto, J.A. Cobos, Comparison of current doubler rectifier and center tapped rectifier for low voltage applications, in *Twenty-First Annual IEEE Applied Power Electronics Conference and Exposition, 2006. APEC '06* (2006), p. 7
18. G.T. Chiang, K. Orikawa, Y. Ohnuma, J.i. Itoh, Improvement of output voltage with svm in three-phase ac to dc isolated matrix converter, in *Industrial Electronics Society, IECON 2013 - 39th Annual Conference of the IEEE* (2013), pp. 4862–4867
19. A. Stupar, T. Friedli, J. Minibock, J.W. Kolar, Towards a 99% efficient three-phase buck-type pfc rectifier for 400-v dc distribution systems. IEEE Trans. Power Electron. **27**, 1732–1744 (2012)
20. H. She, H. Lin, X. Wang, L. Yue, Damped input filter design of matrix converter, in *International Conference on Power Electronics and Drive Systems, 2009. PEDS 2009* (2009), pp. 672–677
21. R. Dubey, P. Agarwal, M. Vasantha, Programmable logic devices for motion control mdash;a review. IEEE Trans. Ind. Electron. **54**, 559–566 (2007)
22. M. Hamouda, H. Blanchette, K. Al-Haddad, F. Fnaiech, An efficient dsp-fpga-based real-time implementation method of svm algorithms for an indirect matrix converter. IEEE Trans. Ind. Electron. **58**, 5024–5031 (2011)
23. Auto transformer rectifier unit. http://www.excelitas.com/Downloads/DTS_ATRU_750.pdf

Chapter 3
A Matrix Based Isolated Three Phase AC–DC Converter

3.1 Introduction

In recent efforts of making aircraft more energy efficient, aircraft-industries are moving towards More Electric Aircraft (MEA). MEA offers several benefits compared to a conventional aircraft system including improved power transmission efficiency, reduced fuel consumption, lesser weight and reduced environmental impact. One of the enabling technologies for MEA is power electronic converter which is required to convert and condition the generated electric power for different aircraft loads [1–3].

In Chap. 2, a non-isolated matrix based AC–DC converter is presented. In this chapter, the non-isolated matrix based AC–DC converter topology is extended for isolated power conversion using high frequency AC transformer. In Sect. 3.2, the brief review of isolated AC–DC converter is presented. The new contributions of the chapter are highlighted in Sect. 3.3. Subsequently, topology and operation of the converter are discussed in Sect. 3.4. Comprehensive steady state analysis and design are carried out in Sect. 3.5. The proposed converter is simulated in MATLAB and simulation results are discussed in Sect. 3.6. In Sect. 3.7, the digital implementation of the proposed SVM based modulation scheme is discussed in details. The hardware implementation of the proposed converter is done in Sect. 3.8 and experimental results are discussed. In Sect. 3.9, the comparative evaluation of the proposed converter with existing isolated three phase AC–DC converter is carried out. Section 3.10 provides the conclusion.

3.2 Brief Review of the Isolated AC–DC Converter

The power generated inside the aircraft is three phase AC with variable frequency (350–800 Hz). Various power converters including AC–DC, AC–AC and DC–DC are employed inside the aircraft for feeding power to different electrical loads. Electrical

© Springer Nature Singapore Pte Ltd. 2018
A. K. Singh, *Analysis and Design of Power Converter Topologies
for Application in Future More Electric Aircraft*, Springer Theses,
https://doi.org/10.1007/978-981-10-8213-9_3

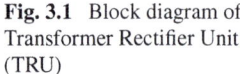

Fig. 3.1 Block diagram of Transformer Rectifier Unit (TRU)

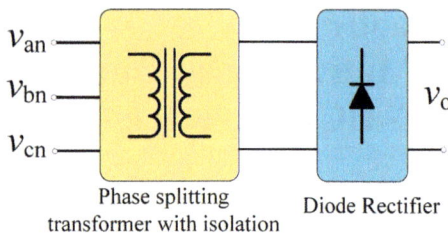

Phase splitting transformer with isolation   Diode Rectifier

isolation is preferred for certain electrical loads such as batteries and other noise-sensitive loads. Power density (output power/weight) is an important consideration for the power electronic converter used in aircraft system. Additionally, the power converters used in aircraft system are often exposed to harsh operating environment. The high emphasis on *power density* and *reliability* specializes the design of power converter required for aircraft system [4–7].

Presently, the aircraft system uses passive Transformer Rectifier Unit (TRU) for AC–DC conversion [8, 9]. The basic schematic of a TRU unit is shown in Fig. 3.1. It consists of a multiphase isolation transformer followed by a diode bridge rectifier. The multiphase transformer is required to limit the Total Harmonic Distortion (THD) within the specifications set by DO-160G standards. The TRU is a passive rectifier and is preferred for its simplicity in design which is crucial to high reliability. However, the passive TRU requires bulky 400 Hz multiphase transformer and does not facilitate controllable power factor and active damping.

The limitation of the passive TRUs can be overcome by active TRUs. An active TRU employs controllable switches and replaces the bulky low frequency transformer with a high frequency transformer. It also has power factor control and active damping capabilities unlike the passive TRU. Basically, two structure are possible for designing an active TRU - 1. Back-to-back isolated rectifier 2. Single-stage isolated rectifier [10–12]. As shown in Fig. 3.2, a back-to-back isolated rectifier is a two-stage converter [13]. In the first stage, three phase AC voltages are rectified to DC voltage whereas in the second stage, an isolated DC–DC converter with a high frequency transformer is used. These two stages are linked with an intermediate DC link capacitor which is usually an electrolytic capacitor. For high power ratings, the DC link capacitor becomes quite bulky and therefore, is not suitable for the applications where high power density is essential. Moreover, the limited life span of the electrolytic capacitor combined with its vulnerability to high temperature reduces the reliability of the overall system. Several single-stage isolated rectifier topologies have been proposed for three phase AC–DC conversion in the literature [14, 15, 15–17]. The single-stage conversion eliminates the bulky DC link capacitor and therefore, promises higher power density and reliability. In single-stage topologies, the three phase line frequency AC voltages are directly converted into single phase high frequency AC voltage. By generating single phase high frequency AC voltage

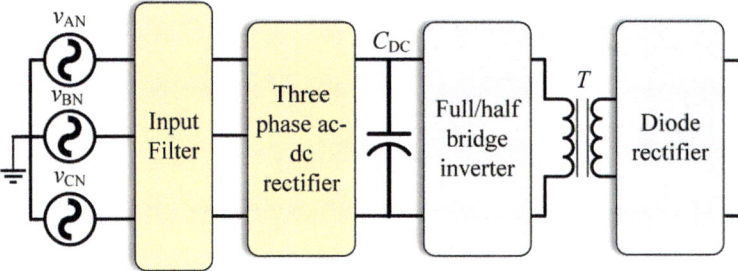

Fig. 3.2 Block diagram of a conventional back to back converter

directly, significant reduction in magnetic and passive elements can be achieved. In [14], a matrix based AC–DC converter for telecommunication application has been proposed. It uses fictitious DC link modulation based on SVM for the single-stage conversion of the three phase AC voltages into DC voltage with electrical isolation. However, it does not address the transformer current commutation issues arising due to unavoidable finite leakage inductance of the high frequency transformer. Moreover, a buck type of rectifier must have dead time between the adjacent switches to avoid the short circuiting of input filter capacitor which is not addressed in most of the papers related to single-stage conversion using the matrix topology. The problem of commutation delay due to the leakage inductance of the high frequency transformer for a three phase matrix based AC–DC converter is highlighted in [18]. In [15], a SPWM based modulation technique has been proposed for the matrix based single-stage AC–DC conversion where only simulation results have been presented to show the feasibility of the converter. However, SVM based modulation schemes are found to be better than SPWM based modulation scheme because of their reduced harmonic content. A number of papers have been published for single-stage conversion using resonant link [19–24]. However, the complexity of control and modulation technique limits the implementation of the converter at lower switching frequencies (≤ 10 kHz). And therefore, even if the switches achieve ZVS, it does not provide significant improvement in power conversion efficiency and power density. In [17], a flyback converter based single-stage converter is presented which requires less number of active switches and provides simpler implementation of the switching scheme. However, the power conversion efficiency of the presented converter is lower due to higher conduction losses in its operation. A matrix-converter-based Inductive Power Transfer (IPT) system, which employs high-speed SiC devices to facilitate the generation of high-frequency current through a single power conversion stage is proposed in [25]. Even though, the proposed converter is implemented with simpler switching scheme, it fails to provide the power quality required in the aircraft systems.

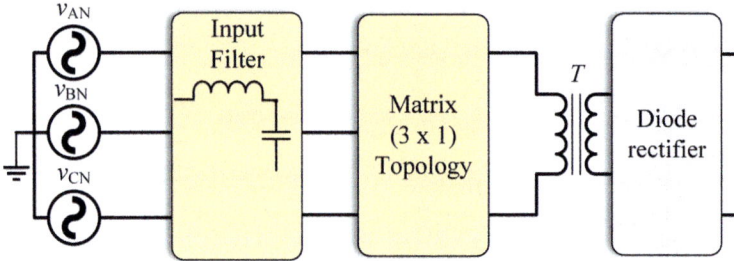

Fig. 3.3 Block diagram of the proposed matrix based AC–DC converter

3.3 New Contribution of the Chapter

In this chapter, a CDR based isolated single-stage three phase AC–DC converter is described as shown in Fig. 3.3. A SVM based modulation scheme is proposed for superior input power quality and high power conversion efficiency. In the proposed modulation scheme, single isolated gate-drive control for each of the bidirectional switches is carried out which essentially simplifies the digital implementation of the proposed switching scheme. Moreover, it also reduces the number of isolated gate drivers required for the matrix switches. The proposed modulation scheme does not require switch body diode commutation and therefore, promises lower switch conduction loss. Since the body diodes of the matrix switches never conduct in the proposed modulation scheme, there are no reverse recovery losses either. Matrix converter is inherently buck type and therefore, the implementation of SVM mandates the use of dead time between the adjacent switches of the two matrix legs to prevent the short circuit of the input filter capacitors. However, the dead time interrupts the transformer primary current and can potentially generate high voltage transients across the matrix switches due to high $\left[\frac{di}{dt}\right]$. The proposed modulation scheme considers the finite leakage inductance of the high frequency transformer and provides lossless current commutation by adding a series capacitor in the primary winding of the high frequency transformer. Consequently, it reduces the switch voltage spikes and improves input current THD. Even though the proposed converter uses one additional bidirectional switch to eliminate the high voltage spikes arising due to the finite leakage inductance of the high frequency transformer, the conduction period of the switches is very small. Moreover, they also turn ON with ZVS and therefore, do not impact the overall power conversion efficiency. In brief, followings are the main contributions of this chapter:

1. proposes an isolated single-stage matrix based three phase AC–DC converter with a CDR for improved power conversion efficiency and power density;
2. proposes a SVM based modulation scheme for directly converting the three phase AC voltages into bipolar single phase high frequency AC voltage with superior input power quality. The proposed scheme require only one gate drive control for one matrix switch and therefore, reduces the control complexity. Moreover, it

does not require switch body diode conduction and therefore, the reverse recovery losses as well as the conduction losses due to the forward voltage drop of the body diodes are eliminated;

3. proposes a novel approach of adding a series capacitor with the leakage inductance of the high frequency transformer which facilitates

   a. soft commutation of the high frequency current from positive to negative value and vice versa

   b. eliminates the duty cycle loss which arises due to the leakage inductance of the high frequency transformer and

4. An efficient digital implementation of the proposed modulation scheme using the combination of DSP and FPGA is carried out for high switching frequency and step-by-step implementation is discussed.

5. Comparative evaluation of the proposed converter with other isolated AC–DC converters in literature is carried out. The effects of mandatory intermediate DC link capacitor in a back to back converter on the overall power density, power conversion efficiency and reliability are discussed.

3.4 Topology and Principle of Operation

Figure 3.4 shows the proposed single-stage matrix based isolated three-phase AC–DC converter with a CDR. It consists of three-phase AC input voltages (v_{an}, v_{bn}, v_{cn}), input filter with inductors (L_a, L_b, L_c) and capacitors (C_a, C_b, C_c), six bidirectional switches (S_1–S_6), leakage inductor (L_r), high frequency isolation transformer (T), CDR diodes (D_1, D_2), two identical output filter inductors (L_{f1}, L_{f2}), output capacitor (C_o) and a constant current load, I_o. A shorting leg has been formed by connecting two back to back switches, S_{w1} and S_{w2} with body diodes D_{s1} and D_{s2}, respectively. A high frequency AC capacitor, C_r is added in series with the leakage inductor, L_r.

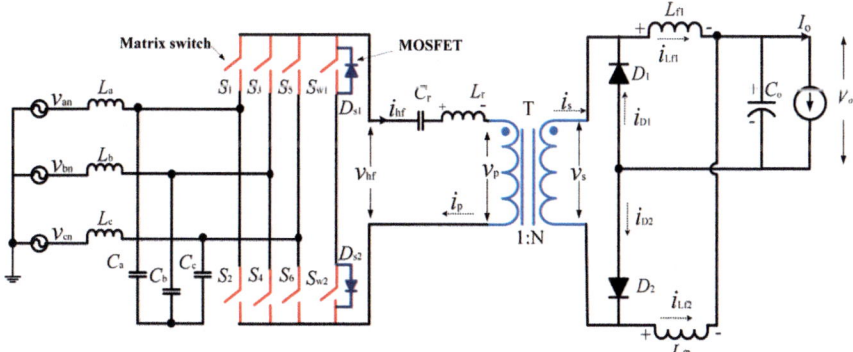

Fig. 3.4 Proposed topology of the Matrix based AC–DC converter

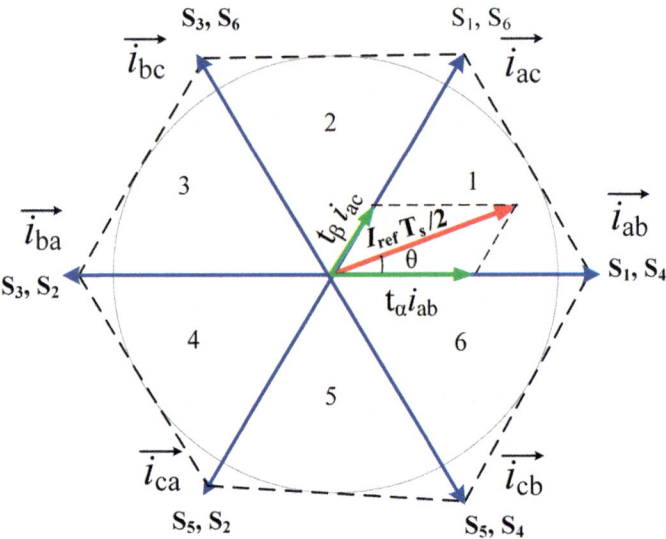

Fig. 3.5 Phaser diagram for SVM

To derive the switching signals for the matrix (3×1) converter, the three-phase input AC voltages shown in (3.1) are divided into six equal sectors as shown in Fig. 3.5.

$$v_{an} = V_m \sin \theta; \ v_{bn} = V_m \sin \left(\theta - \frac{2\pi}{3} \right);$$
$$v_{cn} = V_m \sin \left(\theta + \frac{2\pi}{3} \right) \tag{3.1}$$

During each sector, the current vector, I_{ref} can be synthesized by SVM method. For sector-1, I_{ref} can be written as,

$$I_{ref} \frac{T_s}{2} = i_{ab} t_\alpha + i_{ac} t_\beta \tag{3.2}$$

where, t_α and t_β are the time for which switches S_6 and S_2 are turned ON, respectively. If T_s is the time period of one switching cycle, then t_α and t_β can be derived as

$$t_\alpha = m \frac{T_s}{2} \sin \left(\frac{2\pi}{3} - \theta \right); \ t_\beta = m \frac{T_s}{2} \sin \left(\theta - \frac{\pi}{3} \right);$$
$$t_0 = \frac{T_s}{2} - t_\alpha - t_\beta \tag{3.3}$$

where, θ is the angle of the current vector I_{ref}. For example, θ varies from $\frac{\pi}{3}$ to $\frac{2\pi}{3}$ for sector-1. The time interval, t_o represents the time interval for which all switches are in off state. It is important to note here that the leakage inductance of the high frequency transformer does not allow any open circuit condition in the output of the matrix converter. Therefore, a path for the high frequency AC current should be provided for uninterrupted commutation. The path for the high frequency AC current is provided by turning on the switches S_{w1} and S_{w2}. During this period, because of the series capacitor, C_r, the high frequency AC current changes its direction due to the resonance between L_r and C_r and therefore, provides soft and lossless current commutation. The details of this operation are explained in the modes of operation. To generate high frequency AC output, an inversion signal (U) is used to invert the matrix output voltage after half of the switching cycle ($\frac{T_s}{2}$). The switching sequence of the active and zero states are arranged symmetrically to generate symmetrical balanced bi-polar high Frequency AC output, v_{hf}. For sector-1, the current vectors, $\vec{i_{ab}}$ and $\vec{i_{ac}}$ are used to synthesize the reference current vector, $\vec{I_{ref}}$. For a given modulation index, m, the time durations, t_α and t_β are calculated which are subsequently, arranged in a switching cycle. The following assumptions are made to study the operation and analysis of the converter.

- The filter capacitor voltages are of purely sinusoidal shape and in phase with the three phase input AC voltages;
- reactive currents due to filter capacitors are neglected;
- all the active and passive components are assumed to be ideal;
- magnetizing inductance of the transformer is infinitely large which results in $i_{hf} = i_p$;
- the output DC current, I_o is assumed to be constant. Consequently, the current in each of the CDR inductors, $i_{L_{f1}}$ and $i_{L_{f2}}$ is half of the output DC current, $\frac{I_o}{2}$.

The complete modes of operation of the proposed converter is divided into 14 modes. However, because of the symmetrical operation during positive and negative cycle, the operation for the first seven modes are sufficient for the analysis and design of the proposed converter. The theoretical modes of operation of the proposed converter is shown in Fig. 3.6. For the first 7 modes (mode-1 to mode-7), the inversion signal, $U = 1$, whereas for next 7 modes (mode-8 to mode-14), the inversion signal, $U = 0$.

Mode-1 (*$t_0 \leq t \leq t_1$*): Before the mode-1 starts, there exists a finite dead-time, t_d during which primary current, i_p flows through the shorting leg discharging the switch capacitor, S_{w2}. Therefore, at the start of this mode, the switch, S_{w2} is turned ON with ZVS. During this mode, both switches S_{w1} and S_{w2} remain ON and therefore, provide path for the primary current, i_p to flow without any interruption. The high frequency AC voltage, v_{hf} is zero during this period. Both CDR diodes, D_1 and D_2 conduct and share the output current, I_o. However, during this mode, the current, I_{D1} starts decreasing and the current, I_{D2} starts increasing in such a way that the sum of the two currents is equal to output current, I_o and the difference of the two currents is $\frac{i_p}{N}$. At the end of this mode, the current, i_p becomes zero and thus, both CDR diodes, D_1 and D_2 carries half of the output current, $\frac{I_o}{2}$. The circuit operation during this mode of operation is shown in Fig. 3.7a. The governing equation during this mode is given by,

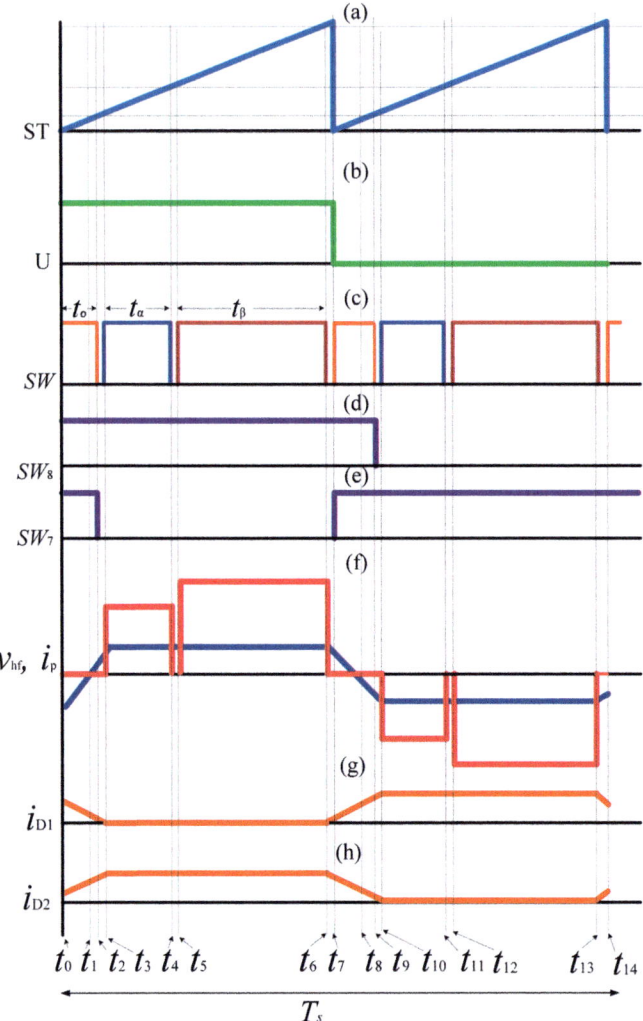

Fig. 3.6 Theoretical modes of operation of the proposed matrix based isolated three phase AC–DC converter. **a** Sawtooth signal, SW. **b** Inversion signal, U. **c** Switching sequence for a switching cycle during sector-1. **d** Switching Signal for shorting leg switch, S_{w2}. **e** Switching Signal for shorting leg switch, S_{w1}. **f** High frequency AC voltage, v_{hf} and current, i_{hf}. **g** Diode D_1 current, i_{D1}. **h** Diode D_2 current, i_{D2}

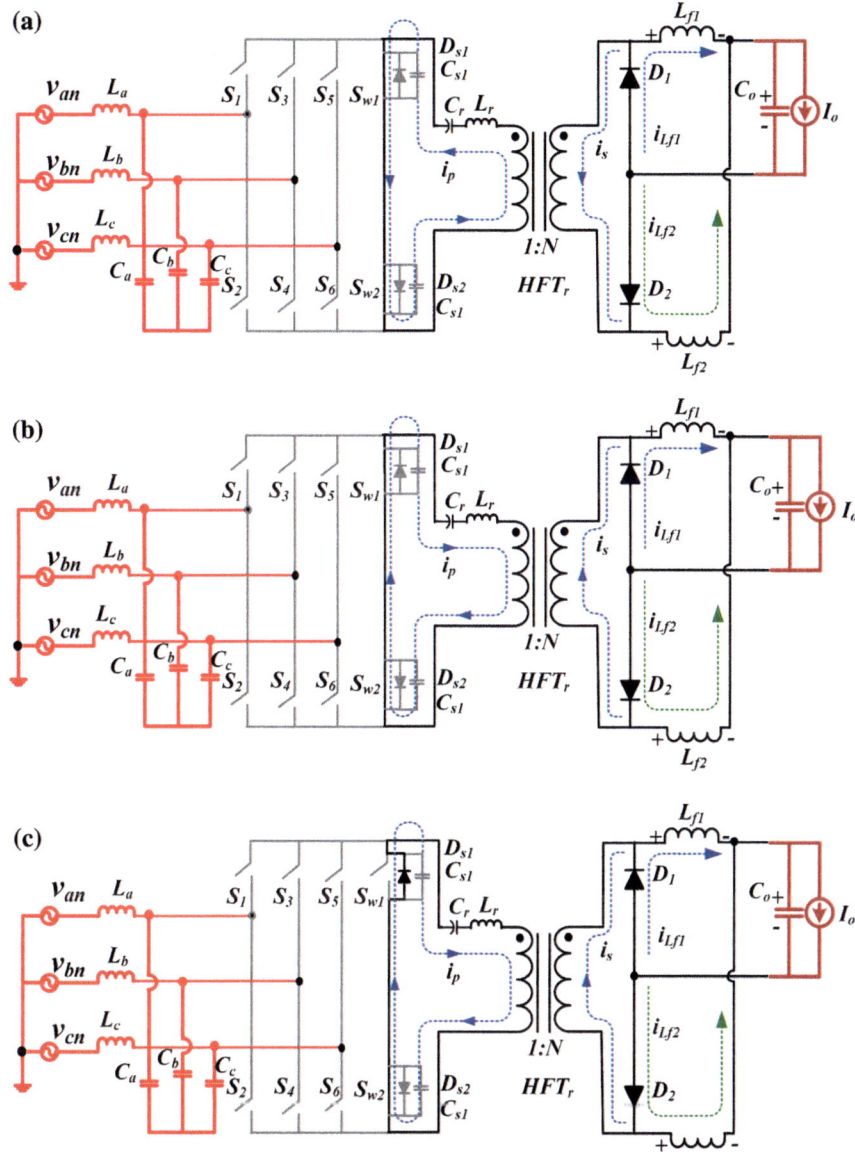

Fig. 3.7 Different modes of operation of the proposed matrix based isolated three phase AC–DC converter. **a** mode-1, **b** mode-2, **c** mode-3

$$v_{hf}(t - t_o) = 0; \tag{3.4}$$

$$L_r C_r \frac{d^2 i_p(t - t_o)}{dt^2} - i_p(t - t_o) = 0; \tag{3.5}$$

$$i_{D1}(t - t_o) = \frac{N I_o - 2 i_p(t - t_o)}{2N}; \tag{3.6}$$

$$i_{D2}(t - t_o) = \frac{N I_o + 2 i_p(t - t_o)}{2N} \tag{3.7}$$

Mode-2 ($t_1 \leq t \leq t_2$): At the start of this mode, the high frequency AC current, i_p is zero. Due to the resonance between L_r and C_r, i_p changes direction and starts increasing in the positive direction. At the CDR side, both diodes D_1 and D_2 conduct. The circuit operation during this mode of operation is shown in Fig. 3.7b. The governing equation during this mode is given as,

$$v_{hf}(t - t_1) = 0; \tag{3.8}$$

$$L_r C_r \frac{d^2 i_p(t - t_1)}{dt^2} - i_p(t - t_1) = 0; \ i_p(t_1) = 0 \tag{3.9}$$

$$i_{D1}(t - t_1) = \frac{N I_o - 2 i_p(t - t_1)}{2N}; \tag{3.10}$$

$$i_{D2}(t - t_1) = \frac{N I_o + 2 i_p(t - t_1)}{2N} \tag{3.11}$$

Mode-3 ($t_2 \leq t \leq t_3$): At the start of this mode, the switch S_{w1} is turned off. The current, i_p starts flowing through the switch body diode, D_{s1} without any interruption. The operation of the CDR remains exactly similar to mode-2. The circuit operation during this mode of operation is shown in Fig. 3.7c. The governing equation during this modes is given as,

$$v_{hf}(t - t_2) = 0; \tag{3.12}$$

$$L_r C_r \frac{d^2 i_p(t - t_2)}{dt^2} - i_p(t - t_2) = 0; \ i_p(t_2) = \frac{N I_o}{2} \tag{3.13}$$

$$i_{D1}(t - t_2) = \frac{N I_o - 2 i_p(t - t_2)}{2N}; \tag{3.14}$$

$$i_{D2}(t - t_2) = \frac{N I_o + 2 i_p(t - t_2)}{2N} \tag{3.15}$$

Mode-4 ($t_3 \leq t \leq t_4$): This mode starts when the switches S_1 and S_6 are turned ON. The matrix switches, S_1 and S_6 remain turned on for t_α duration. During this mode, the high frequency AC current, i_p starts flowing from phase-a to Phase-c giving rise to high frequency AC voltage, $v_{hf} = v_{ac}$. At the CDR side, the diode, D_1 stops conducting. The transformer secondary current, i_s becomes equal to $\frac{I_o}{2}$ and flows through the inductor, L_{f2}. The circuit operation during this mode is given in Fig. 3.8d. The governing equation during this modes is given as,

$$v_{hf}(t - t_3) = v_{ac}; i_p(t - t_3) = \frac{NI_o}{2} \tag{3.16}$$

$$i_{D1}(t - t_3) = 0; i_{D2}(t - t_3) = I_o \tag{3.17}$$

Mode-5 ($t_4 \leq t \leq t_5$): This mode starts when the switch, S_6 is turned OFF and exists for the time duration, t_d. The time duration, t_d is the dead time provided between the switching transition from the switch, S_6 to the switch, S_4 to prevent the shorting of the input filter capacitors. During this mode, the primary current flows through the shorting leg formed by the switch S_{w2} (ON) and the body diode, D_{s1} without having any interruption and therefore, eliminates the voltage spikes caused by open-circuit condition of the matrix output. During this mode, the CDR diodes, D_1 and D_2 conduct. The circuit operation during this mode is given in Fig. 3.8e. The governing equation during this mode is given by,

$$v_{hf}(t - t_4) = 0; \tag{3.18}$$

$$L_r C_r \frac{d^2 i_p(t - t_4)}{dt^2} - i_p(t - t_4) = 0; i_p(t_4) = \frac{NI_o}{2} \tag{3.19}$$

$$i_{D1}(t - t_4) = \frac{NI_o - 2i_p(t - t_4)}{2N}; \tag{3.20}$$

$$i_{D2}(t - t_4) = \frac{NI_o + 2i_p(t - t_4)}{2N} \tag{3.21}$$

As the duration of the dead time, t_d is very small compared to the switching period, T_s, the primary current, i_p is assumed to be constant. With this assumption,

$$i_p(t - t_4) = \frac{NI_o}{2}; i_{D1}(t - t_4) = 0; i_{D2}(t - t_4) = I_o \tag{3.22}$$

Mode-6 ($t_5 \leq t \leq t_6$): This mode starts when the matrix switch S_4 is turned on. The matrix switches, S_1 and S_4 remain turned on for t_β duration. During this mode, the high frequency AC current, i_p starts flowing from phase-a to Phase-b giving rise to high frequency AC voltage, $v_{hf} = v_{ab}$. At the CDR side, the diode, D_1 stops conducting. The transformer secondary current, i_s becomes equal to $\frac{I_o}{2}$ and flows through the inductor, L_{f2}. The circuit operation during this mode is given in Fig. 3.8f. The governing equation during this mode of operation is given as,

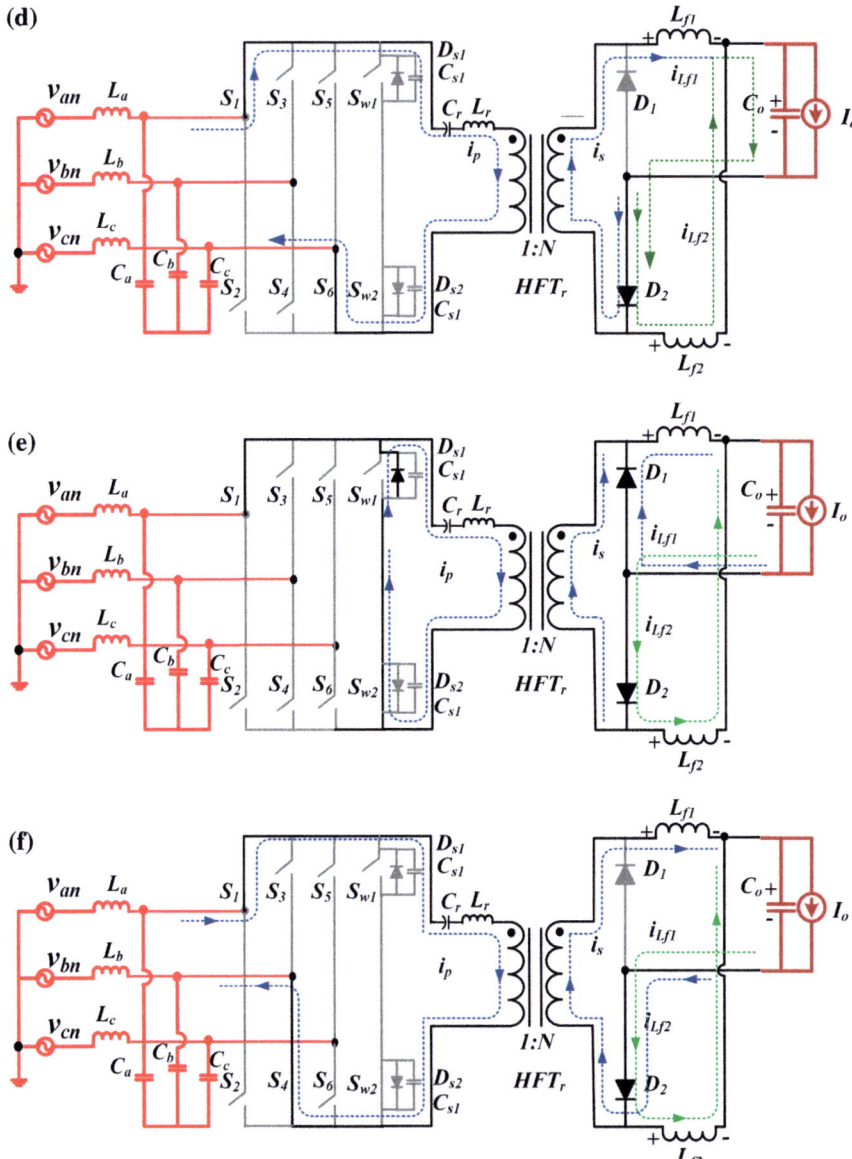

Fig. 3.8 Different modes of operation of the proposed matrix based isolated three phase AC–DC converter. **d** mode-4, **e** mode-5, **f** mode-6

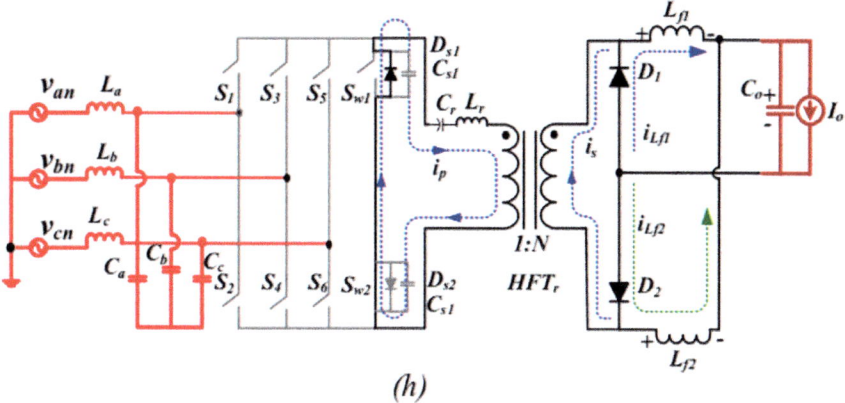

(h)

Fig. 3.9 Different modes of operation of the proposed matrix based isolated three phase AC–DC converter (mode-8)

$$v_{hf}(t - t_5) = v_{ac}; \ i_p(t - t_5) = \frac{N I_o}{2}; \qquad (3.23)$$

$$i_{D1}(t - t_5) = 0; \ i_{D2}(t - t_5) = I_o \qquad (3.24)$$

Mode-7 $(t_6 \le t \le t_7)$: This mode exists for t_d time duration and starts when the switch S_4 is turned off. During this mode, the high frequency AC current, i_p flows through the shorting leg formed by the switch S_{w2} (ON) and the body diode, D_{s1} without having any interruption. This mode ends with the switch S_{w1} turned ON. During this mode, the CDR diodes, D_1 and D_2 conduct. The circuit operation during this mode is given in Fig. 3.9. The governing equation during this mode is given as,

$$v_{hf}(t - t_6) = 0; \qquad (3.25)$$

$$L_r C_r \frac{d^2 i_p(t - t_6)}{dt^2} - i_p(t - t_6) = 0; \ i_p(t_6) = \frac{N I_o}{2} \qquad (3.26)$$

$$i_{D1}(t - t_6) = \frac{N I_o - 2i_p(t - t_6)}{2N}; \qquad (3.27)$$

$$i_{D2}(t - t_6) = \frac{N I_o + 2i_p(t - t_6)}{2N} \qquad (3.28)$$

As the duration of the dead time, t_d is very small compared to the switching period, T_s, the primary current, i_p is assumed to be constant. With this assumption,

$$i_p(t - t_6) = \frac{N I_o}{2}; \ i_{D1}(t - t_6) = 0; \ i_{D2}(t - t_6) = I_o \qquad (3.29)$$

The end of mode-7 completes the half of the switching cycle. Similarly, the next half of the switching cycle is described from mode-8 to mode-14 as shown in Fig. 3.8. The combined 14 modes of operation completes the full operation of the proposed converter.

3.5   Steady State Analysis and Design

Based on the modes of operation described in Sect. 3.4, the steady state analysis and design of the proposed converter are presented in this section. Moreover, the voltage gain of the proposed converter is derived and the effect of leakage inductance on the output voltage is analyzed The voltage and current stresses of the active and passive devices used in the proposed converter are derived. Subsequently, the magnetics design for the proposed converter is carried out. The effect of adding capacitor in series with the primary winding of the high frequency transformer is discussed in details and subsequently, design equation for the series capacitor is derived.

3.5.1   Derivation of the Output Voltage of the Proposed Converter

The output voltage, V_o of the proposed converter can be derived without considering the leakage inductance of the high frequency transformer by using volt-time balance across one of the output filter inductors (L_{f1}, L_{f2}). The voltage across the inductor, L_{f1} is given by $(Nv_p(t) - V_o)$. Based on Fig. 3.6f and neglecting the dead time duration, following volt-time balance equation for the total duration, T_s can be derived,

$$\underbrace{-V_o t_o + \left(Nv_{ab}(t) - V_o\right)t_\alpha + \left(Nv_{ab}(t) - V_o\right)t_\beta}_{\text{Positive half cycle}}$$

$$\underbrace{-V_o(t_o + t_\alpha + t_\beta)}_{\text{Negative half cycle}} = 0 \qquad (3.30)$$

where, $v_{ab} = v_{an} - v_{bn}$ and $v_{ac} = v_{an} - v_{cn}$. Simplifying the (3.30) for sector-1 gives,

$$V_o T_s = N\left(v_{ab}t_\alpha + v_{ac}t_\beta\right) \qquad (3.31)$$

Substituting the values of v_{ab}, v_{ac}, t_α and t_β using (3.1) and (3.3),

$$V_o = \frac{3}{4}mNV_m \qquad (3.32)$$

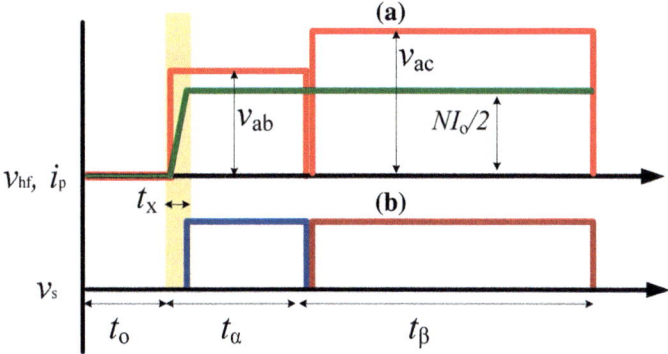

Fig. 3.10 Duty cycle loss due to the leakage inductance of the high frequency transformer. **a** High frequency AC voltage, v_{hf} and current, i_p. **b** Transformer secondary voltage, v_s

where, m is modulation index, V_m is the peak value of the input supply voltage and $1 : N$ is the transformer primary to secondary turns-ratio.

The output voltage described by (3.32) is only valid when there is no leakage inductance. The leakage inductance in a high frequency transformer introduces duty cycle loss as show in Fig. 3.10 which essentially reduces the effective output voltage of the converter. As shown in Fig. 3.10, the time duration t_x results in loss of the effective duty cycle contributing to reduced output voltage of the converter. The time duration, t_x can be expressed as,

$$t_x = \frac{L_r N I_o}{2v_{ab}} \tag{3.33}$$

With t_x into consideration, the output voltage of the proposed converter can be derived by (3.2), (3.3) and (3.33)

$$V_o = \frac{3}{4}mNV_m - \underbrace{\frac{N^2}{2}L_r I_o f_s}_{\Delta V_o} \tag{3.34}$$

It is evident from (3.34) that at higher switching frequency the effect of leakage inductance becomes quite significant which reduces the output DC voltage, V_o. Therefore, at high switching frequency, the transformer leakage inductance should be reduced to minimize duty cycle loss. Figure 3.11 shows duty cycle loss, $\frac{\Delta V_o}{V_o}$ in percentage with respect to switching frequency, f_s variation at different output load, P_o. It is to be noted that at higher power and higher switching frequency, the duty cycle loss becomes very significant. The graph is plotted for $N = 0.56$, $m = 0.7$, $V_o = 48\,\text{V}$, $V_m = 115\sqrt{2}\,\text{V}$ and $L_r = 10\,\mu\text{H}$.

Fig. 3.11 Duty cycle loss variation with respect to switching frequency, f_s at different output power. P_o

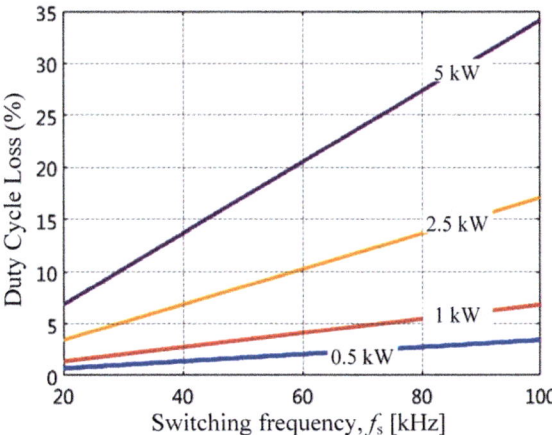

Table 3.1 Voltage stress of the active and passive devices

Parameters	Values
Matrix switches (S_1–S_6; S_{w1}, S_{w2})	$\sqrt{3}\,V_m$
Diodes, D_1, D_2	$\sqrt{3}NV_m$

Table 3.2 Current stress of the active and passive devices

Parameters	Values
Switch rms current (S_1–S_6)	$\frac{I_oN}{2}\sqrt{\frac{m}{\pi}}$
Switch avg current (S_1–S_6)	$\frac{I_omN}{2\pi}$
Diode avg current (D_1, D_2)	$\frac{I_o}{2}$
Diode rms current (D_1, D_2)	$\frac{I_o}{\sqrt{2}}$

3.5.2 Voltage Stress/Current Stress of the Converter

Based on the modes of operation in the Sect. 3.4, the voltage and current stresses of the active and passive devices are calculated and shown in Tables 3.1 and 3.2, respectively.

3.5.3 Magnetics Design

In this subsection the design of magnetics is carried out. The design of magnetics is divided into three parts. 1. Input filter design 2. Transformer design and 3. Output filter design. In this subsection, the design considerations for each of the parts are discussed.

3.5.3.1 Input Filter Design

Unlike the three phase PWM boost rectifiers, the input filter of the proposed converter is LC filter. For high quality sinusoidal input current, capacitors with low ESR, ESL and high current ratings are required. However, a large value of the input capacitor results in capacitive reactive power leading to poor power factor, particularly at low output power. The reactive power in the capacitor is given by,

$$Q_i = 3\omega_i C_i V_{i,rms}^2 \tag{3.35}$$

where, $\omega_i = 2\pi f_i$; C_i is the capacitance value of the input filter capacitor ($C_i = C_a = C_b = C_c$) and $V_{i,rms}$ is the rms value of the input AC voltages.

The input filter capacitor can be designed via the maximum peak-to-peak value of the filter capacitor voltage ripple which is derived as,

$$\Delta V_{C_i,pp} = \frac{\sqrt{3}N I_o}{32 C_i f_s} \tag{3.36}$$

Further, the reactive power Q_i is limited to 5–10% of the rated power in order to ensure high power factor even in case of low load applications. The design objective of the input filter for the proposed converter is to limit the input current THD below 5% and to achieve unity power factor at full load operation (500 W). Another criterion is to limit the maximum peak-to-peak value of the filter capacitor voltage ripple to be less than 5% of the peak of the input line voltage. Moreover, damping is also needed to reduce the input filter LC oscillations. The detailed design of input LC filter is discussed in Chap. 2, Sect. 2.5.5.

3.5.3.2 High Frequency Transformer Design

The design of the high frequency transformer is carried out by assuming the value of, $m = 0.7$ as it provides sufficient space for duty loss and duty cycle overshoots during transients. With this value of m, the first iteration of transformer turns-ratio is calculated using (3.32). The number of turns in the primary winding of the transformer is chosen in such a way that the maximum flux density does not exceed the saturation value of the magnetic core. Ferrite cores offer high performance characteristics for the design of high frequency transformers. It should be noted that the use of CDR reduces the primary to secondary turns ratio by half resulting in smaller window area and reduced copper loss.

Ferrite material based ETD-59 core (3C95 grade) is used and design of magnetics is carried out in such a way that the maximum magnetic flux intensity, B_m remains below 0.25 T. Therefore, for the proper design of the magnetics, the relationship between $B_{m,max}$ and input peak voltage, V_m is derived. Figure 3.12 shows the transformer primary voltage, v_p and the magnetizing current, i_m.

Fig. 3.12 a Transformer primary voltage, v_p. **b** Magnetizing current, i_m

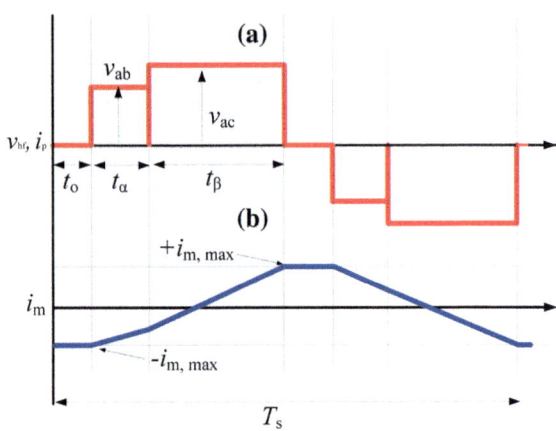

$$I_{m,max} = -I_{m,max} + \frac{v_{ab}t_\alpha}{L_m} + \frac{v_{ac}t_\beta}{L_m} \tag{3.37}$$

$$2I_{m,max} = \frac{1}{L_m}(v_{ab}t_\alpha + v_{ac}t_\beta) \tag{3.38}$$

Above equation can be simplified using, $N_p\phi_{max} = L_m I_{m,max}$; $v_{ab}t_\alpha + v_{ac}t_\beta = \frac{3}{2}m$ $V_m \frac{T_s}{2}$ where, N_p is number of turns in the primary winding of the high frequency transformer; L_m is the magnetizing inductance. Simplifying the above equation results in,

$$V_m = \frac{8}{3}\frac{N_p\phi_{max}}{mT_s} \tag{3.39}$$

Substituting ϕ_{max} with $B_{m,max}A_c$ and $T_s = \frac{1}{f_s}$, (3.39) can be further simplified to,

$$\frac{V_m}{N_p} = \frac{8}{3}\frac{B_{m,max}A_c f_s}{m} \tag{3.40}$$

where, A_c is the cross section area of the magnetic core. For the worst case design, the modulation index m is taken to be maximum ($m = 1$) for the design.

Assuming the dead time t_d to be negligible, the rms value of the transformer primary current, $I_{p,rms}$ can be derived from the modes of operation provided in Sect. 3.4. The derivation of I_{prms} is a two step process. In the first step, the rms value of the primary transformer current. i_p is calculated for a switching cycle which is a function of θ. In the second step, the $I_{p,rms}$ is calculated for one sector duration.

$$I_{p,rms}(\theta) = \frac{NI_o}{2}\sqrt{m\cos\left(\theta - \frac{\pi}{3}\right)} \tag{3.41}$$

where, $I_{p,rms}(\theta)$ is the *rms* value of the high frequency AC current for a switching cycle. Consequently, $I_{p,rms}$ can be derived as follows,

$$I_{p,rms} = \sqrt{\int_0^{\frac{\pi}{3}} \frac{(I_{p,rms}\theta)^2 d\theta}{\frac{\pi}{3}}} = \frac{\sqrt{3}N I_o}{2}\sqrt{\frac{m}{\pi}} \tag{3.42}$$

Similarly, the *rms* value of the transformer secondary current, $I_{s,rms}$ is given as,

$$I_{s,rms} = \frac{I_{p,rms}}{N} = \frac{\sqrt{3}I_o}{2}\sqrt{\frac{m}{\pi}} \tag{3.43}$$

Similarly, the rms value of the high frequency AC voltage, $V_{hf,rms}$ can be derived as,

$$V_{hf,rms} = V_m\sqrt{\frac{3m}{\pi}} \tag{3.44}$$

3.5.3.3 Output Filter Deign

In the output of the proposed converter, a CDR is used which consists of two equal inductors, L_{f1} and L_{f2} and one output capacitor, C_o. The design of output filter inductors, L_{f1} and L_{f2} is carried out in such a way that the peak-to-peak value of the inductor current ripple, $\Delta i_{L,pp,max}$ is limited to a given value. The peak-to-peak value of the inductor current ripple can be calculated by,

$$\Delta i_{L,pp} = \frac{V_o}{L_{f1}}\left(\frac{T_s + t_o}{2}\right) \tag{3.45}$$

where, T_s is one switching period and t_o denotes the total zero period during one switching cycle. With this the output inductor can be selected according to,

$$L_{f1} \geq \frac{V_o}{2\Delta i_{L,pp,max} f_s}\left(1 - \frac{\sqrt{3}m_{min}}{4}\right) \tag{3.46}$$

where, m_{min} is the minimum modulation index and f_s is the switching frequency of the converter. As both of the inductors, L_{f1} and L_{f2} are similar, the design equations are only derived for L_{f1}. The output current is sum of the currents in inductors, L_{f1} and L_{f2}. The two inductors, L_{f1} and L_{f2} behave like interleaved inductors and essentially, reduces the peak-to-peak current ripple in the output current, I_o.

The output capacitor, C_o is selected in order to limit the peak-to-peak value of the output voltage ripple $\Delta v_{C_o,pp,max}$ to a given value. The output capacitor value is given by,

$$C_o \geq \frac{V_o}{32\Delta v_{C_o,pp,max} L_{f1} f_s^2}\left(1 - \frac{\sqrt{3}m_{min}}{4}\right) \tag{3.47}$$

3.5.4 Effect of Adding Series Capacitor and Its Design Consideration

The objective of adding the capacitor, C_r in series with the leakage inductor, L_r is to change the direction of the high frequency AC current, i_p by creating resonance between L_r and C_r during the overlap period of the switches S_{w1} and S_{w2}. By changing the direction of the current, lossless commutation is facilitated contributing to reduced voltage spikes and power loss.

3.5.4.1   Soft Commutation of the High Frequency AC Current

Figure 3.13 shows the sequence of the current commutation in the proposed converter. Figure 3.13a shows the equivalent circuit model just before mode-1 where as Fig. 3.13b–d show the equivalent simplified circuit model for mode-1 and mode-2 and mode-3 respectively.

Before, the start of mode-1 Fig. 3.13a, the current, $i_p = \frac{NI_o}{2}$ flows in the negative direction through the diode, D_{s2}. During mode-1, both of the switches, S_{w1} and S_{w2} are ON and the current, i_p resonates with the leakage inductor, L_r and the series capacitor, C_r as per (3.4). In mode-2, i_p changes its direction and becomes positive. Once i_p becomes equal to $\frac{I_o}{2N}$, mode-3 starts and i_p flows through the diode, D_{s1}. Thus, the current changes direction from negative to positive without any interruption. In the absence of series capacitor, the change of current from positive to negative happens suddenly as shown in Fig. 3.14a which can generate high voltage spikes due to the leakage inductance of the high frequency transformer. By adding an appropriate value of series capacitor with the leakage inductor, the transformer primary current, i_p is shaped as shown in Fig. 3.14b and therefore, the high voltage spikes are eliminated.

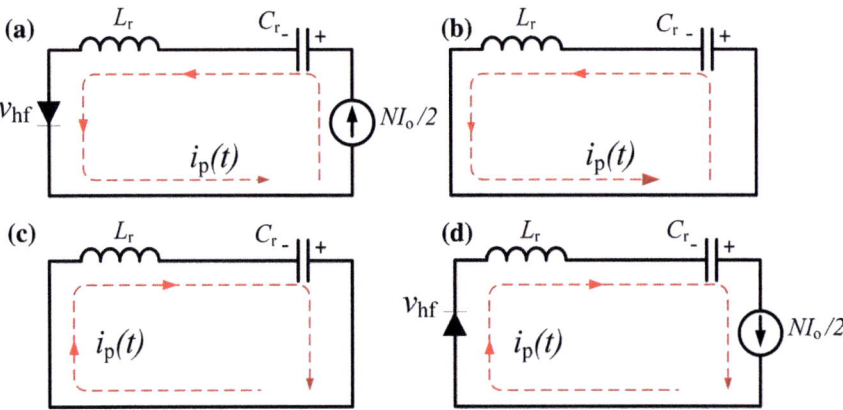

Fig. 3.13 Sequence of the current commutation after adding the series capacitor in the proposed converter. The sequence is *a-b-c-d*

Fig. 3.14 a Sudden change in current in absence of series capacitor. b Soft commutation of current in presence of series capacitor

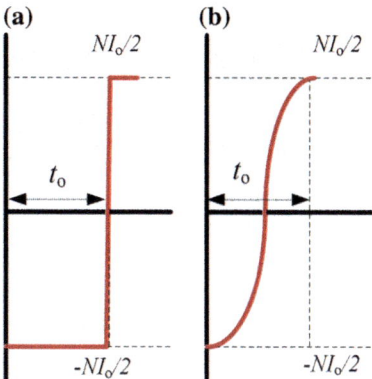

From the Fig. 3.14, the time duration, t_o should be equal to the half of the resonant period which gives,

$$t_o = \pi \sqrt{L_r C_r} \tag{3.48}$$

The duration of t_o varies within a sector and therefore, the minimum value of t_o for a given sector should be equal to the resonant period. The $(t_o)_{min}$ is given as,

$$(t_o)_{min} = \frac{T_s}{2}(1 - m) \tag{3.49}$$

3.5.4.2 Minimization of Duty Cycle Loss

Another advantage of adding capacitor in series with the transformer primary winding is the reduction in duty cycle loss. The duty cycle loss essentially arises as the current, i_p rises from zero to its full magnitude after the high frequency AC voltage appears as shown in Fig. 3.10. By adding an appropriate value of the series capacitor C_r, the current, i_p is made almost equal to its full magnitude $(\frac{I_o}{2N})$ before the high frequency AC voltage, v_{hf} appears by creating resonance between L_r and C_r and therefore, the duty cycle loss in the proposed converter is minimized as shown in Fig. 3.15.

3.5.4.3 Design Consideration of L_r and C_r

The first goal while designing the high frequency transformer is to minimize the leakage inductance as less as possible. However, the presence of leakage inductance is unavoidable in a high frequency transformer and it especially increases for larger turns-ratio. In the proposed converter, the use of CDR in the output side reduces the

Fig. 3.15 Minimizing the
duty cycle loss by adding a
series capacitor with the high
frequency transformer. **a**
High frequency AC voltage,
v_{hf} and current, i_p. **b** voltage
across series capacitor, v_{cr}. **c**
Transformer secondary
voltage, v_s

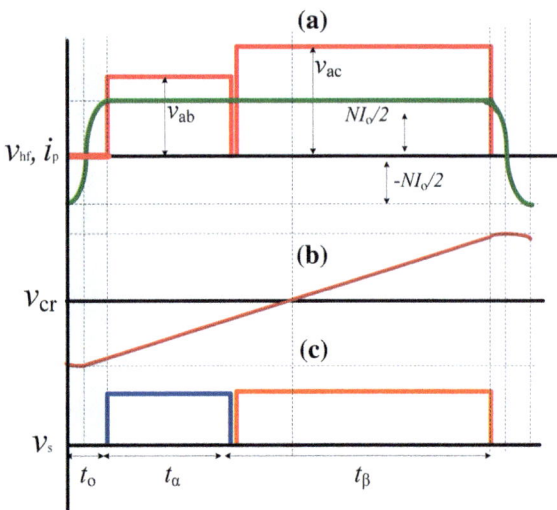

transformer turns-ratio by half contributing to reduced leakage inductance. Once the
leakage inductance of the transformer is known, the value of the series capacitor, C_r
is calculated based on (3.48) and (3.49). For the modulation index, m and the leakage
inductance, L_r, the series capacitor, C_r is derived as,

$$C_r = \frac{1}{4\pi^2 L_r} T_s^2 (1 - m)^2 \tag{3.50}$$

3.6 Simulation of the Proposed Converter

Simulation of the proposed converter is carried out using MATLAB 2015b software.
A comprehensive simulation model is developed in simulink and the performance of
the converter is checked. The specifications of the converter are shown in Table 3.3.

Based on the design equations derived in Sect. 3.5, the various parameters of the
converter are calculated. The input filter is designed with $L_a = L_b = L_c = 200\,\mu\text{H}$,
$C_a = C_b = C_c = 1.2\,\mu\text{F}$. The switching frequency of the converter is chosen to be

Table 3.3 Specifications of
the example converter

Parameters	Values
Input voltage, v_{abc}	115 $V_{ac}(rms)$, 400 Hz
Output voltage, V_o	48 VDC
Switching frequency, f_s	40 kHz
Output power, P_o	500 W

40 kHz as it provides good compromise between the switching loss and the size of passive elements. At 40 kHz switching frequency, the other passive elements such as output filter inductors, $L_{f1} = L_{f2} = 1.2$ mH, $C_o = 400\,\mu F$ are chosen. An ideal transformer with turns-ratio 1 : 0.6 is used in simulation. Further, an external inductor of 10 μH is added in series with the primary winding of the transformer to emulate the leakage inductance, L_r of the high frequency transformer for simulation. The design of resonant capacitor, C_r is carried out based on (3.50) for $m = 0.7$ and $f_s = 40$ kHz. The value of C_r is found to be 0.6 μF which is added in series with the leakage inductance of the high frequency transformer it should be noted that the dead time (t_d) provided between the two adjacent switch transitions become significant with respect to the zero period, t_o and therefore, is considered while designing the series capacitor, C_r.

To validate the analysis carried out in Sect. 3.5, the *rms* and *avg* values of the currents in the matrix switches and the CDR diodes are found both by analysis and digital simulation. Further, the *rms* values of the current in the primary and secondary winding of the high frequency transformer are obtained analytically and compared with the simulation results. The comparison of results is illustrated in Table 3.4. The two solutions show excellent agreement with each other. The simulation results of the proposed converter for the given specifications shown in Table 3.3 are shown in Figs. 3.16, 3.17, 3.18, 3.19, 3.20, 3.21, 3.22, 3.23 and 3.24. Figure 3.16a shows the input three phase AC voltages, v_{an}, v_{bn} and v_{cn}. The input three phase voltages are converted into single phase high frequency AC voltage using the matrix (3×1) topology. Figure 3.16b shows the high frequency AC voltage, v_{hf} which is the output of the matrix converter. The high frequency AC voltage, v_{hf} is processed using a high frequency transformer. The secondary voltage of the high frequency transformer, v_s is shown in Fig. 3.16c. Figure 3.17a shows the symmetrical bipolar high frequency AC voltage, v_{hf}. As shown in Fig. 3.17a, there is a finite amount of dead time,

Table 3.4 Comparison of the results obtained by analytical solution and digital simulation

Parameters	Analytical (A)	Digital (A)
Switch current (Avg)	0.694	0.681
Switch current (rms)	1.47	1.446
Diode current (Avg)	5.20	5.19
Diode current (rms)	7.36	7.27
Transformer primary current (rms)	2.55	2.73
Transformer secondary current (rms)	4.25	4.75

Fig. 3.16 **a** Input three phase voltages, v_{an}, v_{bn} and v_{cn} (V). **b** high frequency AC voltages, v_{hf} (V). **c** Secondary voltage of the high frequency transformer winding, v_s (V)

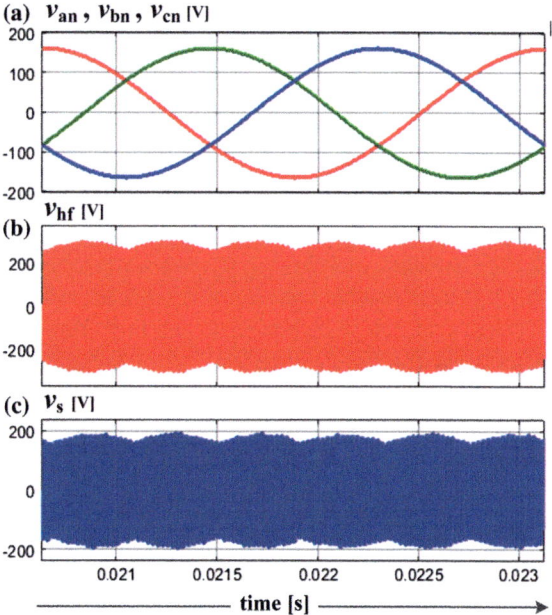

t_d provided between adjacent switches to avoid any short-circuiting of input filter capacitors. Figure 3.17b shows high frequency AC current, i_{hf} which is also input to the primary winding of the high frequency transformer, i_p. Figure 3.17c shows the voltage across the series capacitor, C_r which is added for the soft commutation of the high frequency AC current. Figure 3.18a shows the simulated output DC voltage, V_o. The currents in the output filter inductors, L_{f1} and L_{f2} are shown in Fig. 3.18b, c respectively. The output current, I_o is shown in Fig. 3.18d. It is to be noted that the output current, I_o is the sum of the output filter inductor currents, i_{Lf1} and i_{Lf2}. As the current ripple in the two output inductors are phase shifted by 180°, the output current ripple is reduced in magnitude and doubled in frequency contributing to smaller size of the output filter capacitor. Figure 3.19 shows the input phase voltage, v_{an} and phase current, i_a. The displacement power factor is found to be almost unity. Figure 3.20 shows the input three phase currents, i_a, i_b and i_c at full load. The currents are symmetrical and balanced. The THD of input phase-a current is calculated using MATLAB and plotted in Fig. 3.21. The THD of the current is found to be 2.59%.

The effect of leakage inductor on the duty cycle loss is simulated in MATLAB at different output power. It is found that at higher power, the duty cycle loss becomes quite significant. Figure 3.22 shows the simulated duty cycle loss and theoretical duty cycle loss. Both have been found in excellent agreement with each other. Figure 3.23

Fig. 3.17 **a** High frequency AC voltage, v_{hf} (V). **b** High frequency AC current, i_{hf} which is also the transformer primary current, i_p (A). **c** Voltage across the series capacitor, v_{cr} (V)

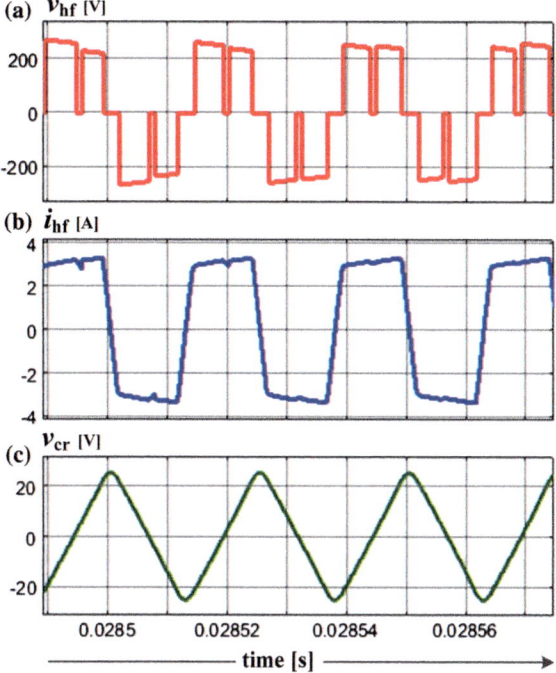

shows the effect of adding an appropriate value of series capacitor in the primary winding of the high frequency transformer. Figure 3.23a shows v_{hf} and i_{hf} without the series capacitor, C_r. It is to be noted that the change of i_{hf} is not smooth resulting in high $\frac{di}{dt}$. Moreover, it also results in a high voltage spike in the high frequency AC voltage, v_{hf} as shown in Fig. 3.23a. Figure 3.23b shows the high frequency AC voltage, v_{hf} and high frequency AC current, i_{hf} when the series capacitor, C_r is added. Due to resonance between L_r and C_r, i_{hf} changes smoothly from negative to positive value during the zero period, t_o. Figure 3.24 shows the effect of adding the series capacitor in minimizing the duty cycle loss of the proposed converter. In the absence of the series capacitor, the high frequency current, i_{hf} starts from zero resulting in duty cycle loss as shown in Fig. 3.24a. Due to resonance between L_r and C_r, the current changes from $-\frac{NI_o}{2}$ to $+\frac{NI_o}{2}$ during the zero period, t_o which results in zero duty cycle loss as shown in Fig. 3.24b. Thus, the results obtained through digital simulation of the proposed converter validate the analysis and design carried out in Sects. 3.4 and 3.5.

Fig. 3.18 **a** Output DC voltage, V_o (V). **b** Current in the output filter inductor, i_{Lf1} (A). **c** Current in the output filter inductor, i_{Lf2} (A). **d** Output DC current, I_o (A)

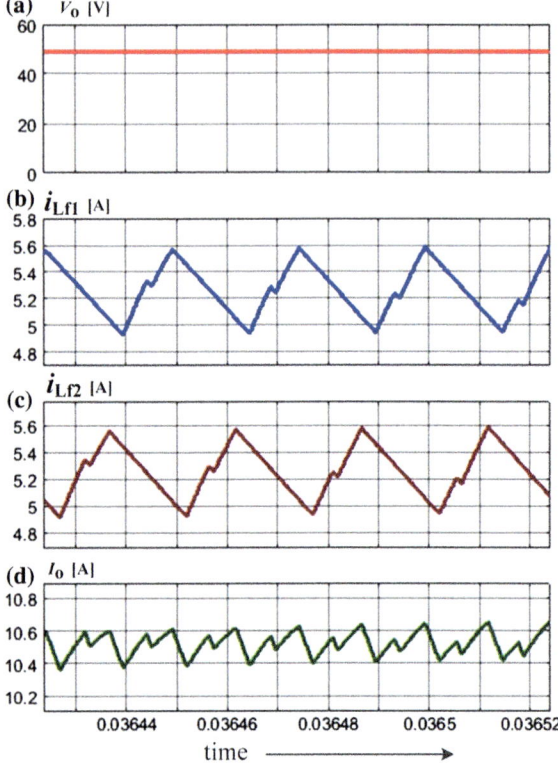

Fig. 3.19 Input phase-a voltage, v_{an} (V) and phase-a current, i_a (A). The displacement power factor is found to be almost unity

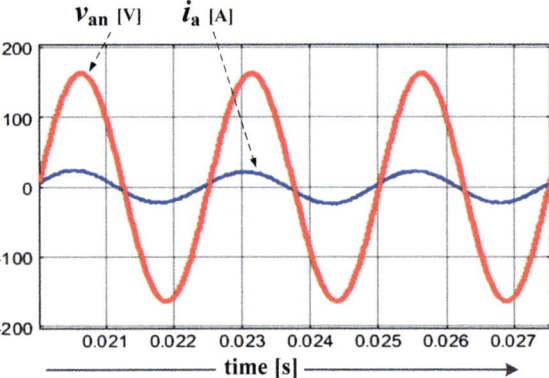

Fig. 3.20 Three phase input current, i_a (A), i_b (A) and i_c (A) at full load of 500 W

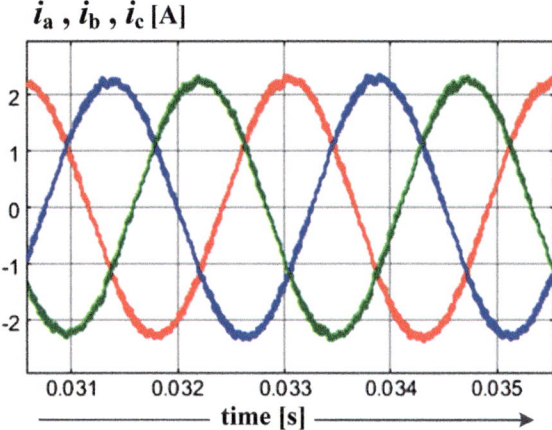

Fig. 3.21 THD of phase-a current, i_a for full load operation

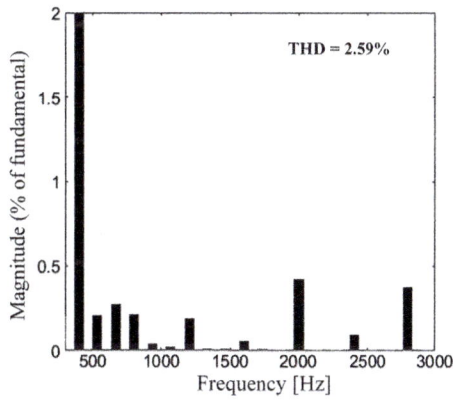

Fig. 3.22 Simulated and theoretical duty loss for different output power at 40 kHz switching frequency. **a** Simulated duty cycle loss in %. **b** Theoretical duty cycle loss in %

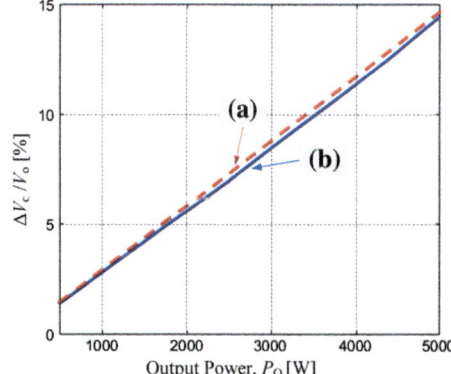

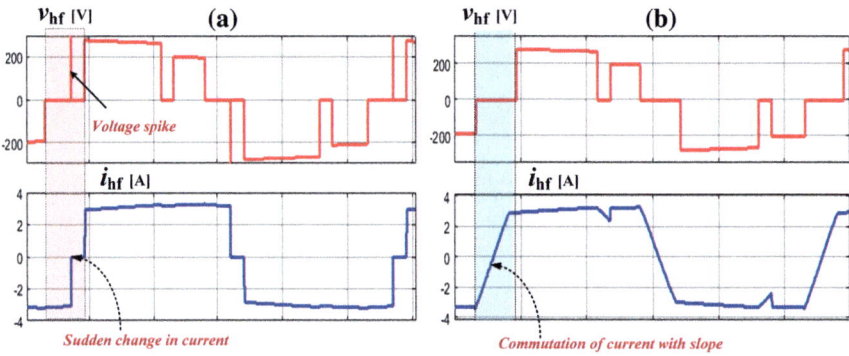

Fig. 3.23 Soft commutation of the high frequency AC current, i_{hf} by adding a series capacitor with the leakage inductance of the high frequency transformer. **a** Without series capacitor. **b** With series capacitor

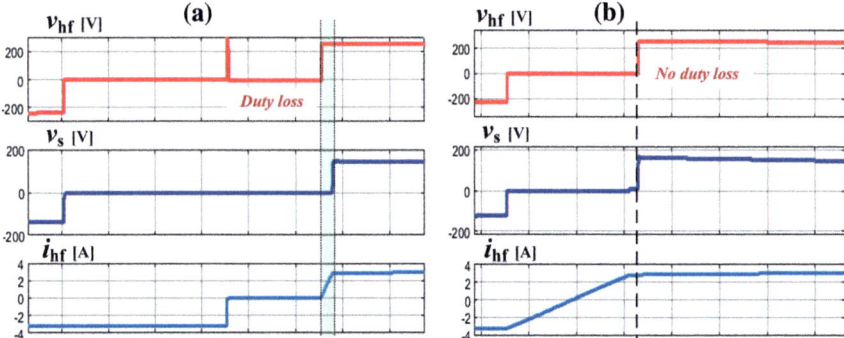

Fig. 3.24 Minimizing the duty loss by adding a series capacitor with the leakage inductance of the high frequency transformer. **a** Without series capacitor. **b** With series capacitor

3.7 Digital Implementation of the Proposed SVM Based Switching Scheme for the Proposed AC–DC Converter

In this section, an efficient real time implementation method of the proposed SVM based modulation scheme is discussed in details. To digitally implement the proposed modulation scheme, a combination of DSP and FPGA is used. Figure 3.25 shows the block diagram of the overall digital implementation. The three phase AC inputs, v_{an}, v_{bn} and v_{cn} as well as modulation index, m are fed to the digital controller which are subsequently processed using DSP and FPGA. The output of the digital controller are eight switching signals which control the eight switches of the proposed matrix based topology as shown in Fig. 3.4. It is worth noticing that in the present implementation the interaction between DSP and FPGA is one-way which reduces

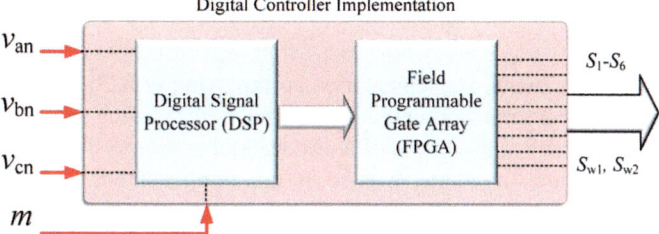

Fig. 3.25 Block diagram of the overall digital control implementation

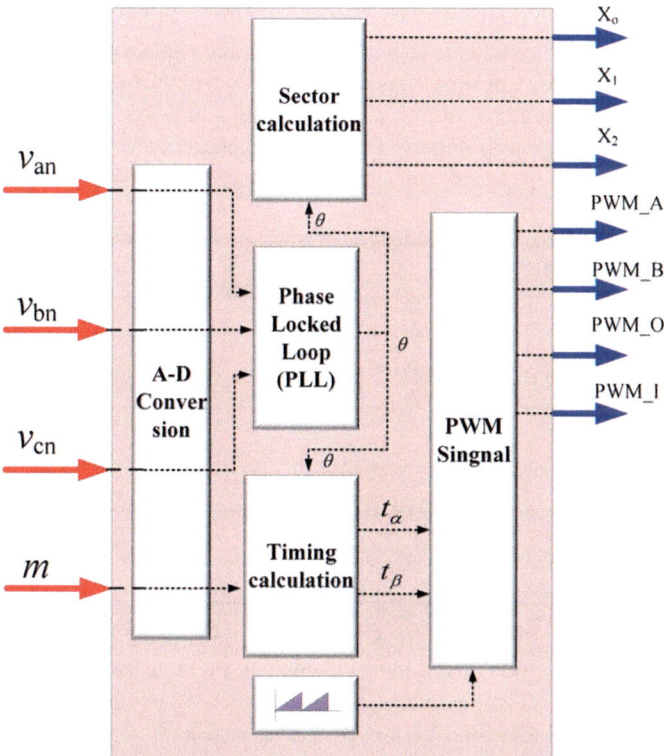

Fig. 3.26 Block diagram of the DSP implementation

implementation complexity. The outputs of the DSP are inputs to the FPGA. The digital implementation of the proposed SVM based modulation scheme is divided into two subsections. The first subsection deals with DSP part of the implementation. In the second subsection, FPGA implementation is discussed.

3.7.1 DSP Implementation

The DSP part of the digital implementation is shown in Fig. 3.26. The sensed three phase AC input voltage and modulation index is fed to the Analogue to Digital (A-D) converter of DSP. Further, the output of A-D converter is fed to a Phase Locked Loop (PLL) which generates the angle, θ. Here, in this implementation, d-q based PLL Implementation is carried out. Based on angle, θ, the three phase input AC voltage is divided into six sectors. Each of the sectors is assigned by 3-bits of digital signal, X_o, X_1, X_2. The timing calculation block calculates t_α and t_β duration based on angle, θ and modulation index, m. Finally, four PWM signals, PWM_A, PWM_B, PWM_O, PWM_I are generated. Figure 3.27 shows the waveforms of these 4 PWM signals. It is to noted that a dead time, t_d is provided between the signals to avoid short-circuit of input filter capacitors during transition. The frequency of the PWM signals are determined by the frequency of the sawtooth signal. Table 3.5 shows the sector and timing calculation for the DSP part of the implementation.

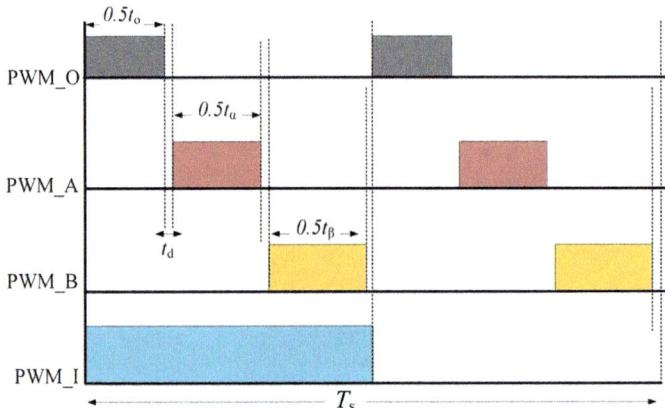

Fig. 3.27 Waveforms of PWM signals, PWM_A, PWM_B, PWM_O, PWM_I

Table 3.5 Sector and timing calculation for DSP implementation

θ	Sector	X_2	X_1	X_1	t_α	t_β
$0 < \theta < \frac{\pi}{3}$	6	1	0	0	$mT_s \sin\theta$	$mT_s \sin(\frac{\pi}{3} - \theta)$
$\frac{\pi}{3} < \theta < \frac{2\pi}{3}$	1	0	0	0	$mT_s \sin(\theta - \frac{\pi}{3})$	$mT_s \sin(\frac{2\pi}{3} - \theta)$
$\frac{2\pi}{3} < \theta < \pi$	2	0	0	1	$mT_s \sin(\theta - \frac{2\pi}{3})$	$mT_s \sin(\pi - \theta)$
$\pi < \theta < \frac{4\pi}{3}$	3	0	1	1	$mT_s \sin(\theta - \pi)$	$mT_s \sin(\frac{4\pi}{3} - \theta)$
$\frac{4\pi}{3} < \theta < \frac{5\pi}{3}$	4	1	1	1	$mT_s \sin(\theta - \frac{4\pi}{3})$	$mT_s \sin(\frac{5\pi}{3} - \theta)$
$\frac{5\pi}{3} < \theta < 2\pi$	5	1	0	1	$mT_s \sin(\theta - \frac{5\pi}{3})$	$mT_s \sin(2\pi - \theta)$

3.7.2 FPGA Implementation

The PWM and sector signals from DSP are fed to the FPGA as shown in Fig. 3.28. The FPGA is programmed for logical operations over these signals and thus, provides the final switching signals for all the matrix switches. The truth tables for the logical operation are shown in Tables 3.6 and 3.7. It is to be noted here that in the proposed implementation, the role of FPGA is only to perform logical operation and therefore, it can be replaced with more cost-effective Programmable Logic Devices (PLD). The switching signals S_{w1} and S_{w2} are generated by PWM_O and PWM_I as shown below.

$S_{w1} = \text{OR (PWM\_O, PWM\_I)}$
$S_{w2} = \text{OR (PWM\_O, NOT(PWM\_I))}$

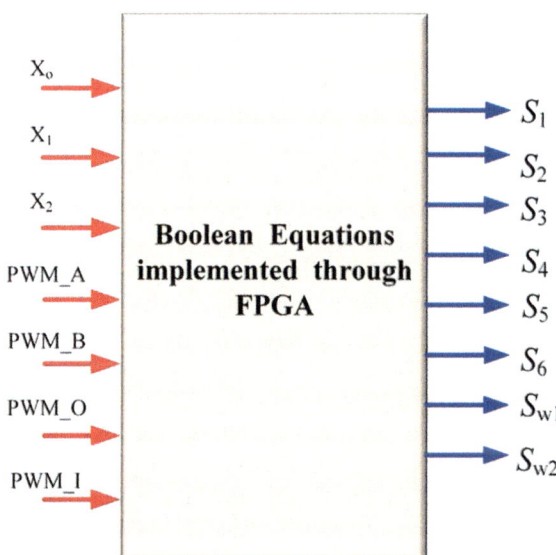

Fig. 3.28 Block diagram of the FPGA implementation

Table 3.6 FPGA truth table for PWM_I = 1

X_2	X_1	X_0	S_1	S_2	S_3	S_4	S_5	S_6
0	0	0	1	0	0	PWM_B	0	PWM_A
0	0	1	PWM_B	0	PWM_A	0	0	1
0	1	1	o	PWM_A	1	0	0	PWM_B
1	1	1	o	1	PWM_B	o	PWM_A	o
1	0	1	0	PWM_B	0	PWM_A	1	0
1	0	0	PWM_A	0	0	1	PWM_B	0

Table 3.7 FPGA truth table for PWM_I = 0

X_2	X_1	X_0	S_1	S_2	S_3	S_4	S_5	S_6
0	0	0	o	1	PWM_B	0	PWM_A	o
0	0	1	0	PWM_B	0	PWM_A	1	0
0	1	1	PWM_A	0	0	1	PWM_B	0
1	1	1	1	0	0	PWM_B	0	PWM_A
1	0	1	PWM_B	0	PWM_A	0	0	1
1	0	0	0	PWM_A	1	0	0	PWM_B

3.8 Experimental Verification

3.8.1 Experimental Results

In the input of the proposed converter, a three phase LC filter is used. Each of the matrix switches (S_1–S_6) is realized using two back to back connected MOSFETs. Moreover, a series capacitor is added in the primary of the high frequency transformer which forms a series resonant tank with the leakage inductor of the high frequency transformer. The output of the high frequency transformer is processed using a CDR circuit. A resistive load, R_o of 4.6 Ω is used to test the converter at full load. A three phase programmable voltage source is used for providing input supply voltage to the proposed converter. All of the devices used to implement the topology shown in Fig. 3.4 are described in Table 3.8.

The three phase input voltages are sensed and given to analogue-to-digital converter of a Digital Signal Processor (DSP). A Texas Instrument DSP, TMDSCN28335 is used for generating the switching signals. A $d-q$ based Phase Locked Loop (PLL) is implemented inside the DSP. The DSP generates the PWM signals as well as three digital signals for the six sectors of the three phase input voltages which are further processed using a ALTERA QUARTUS Field Programmable Gate Array (FPGA) to generate the final switching signals. The generated switching signals are given to the isolated gate drivers (CREE gate driver) to control the matrix switches.

Figure 3.29 shows input phase-a voltage, v_{an}, high frequency AC voltage, v_{hf} and output DC voltage, V_o. The three phase AC voltages are converted into high frequency AC voltage in a single-stage using the matrix topology. Subsequently, a high frequency transformer is used for galvanic isolation as well as necessary voltage step down which is further rectified using a CDR circuit to generate output DC voltage. Figure 3.30 shows the high frequency AC voltage, v_{hf} and the transformer secondary voltage, v_s. The high frequency transformer steps down the input high frequency AC voltage to obtain the required output DC voltage.

The high frequency AC voltage, v_{hf}, the transformer secondary voltage, v_s and the high frequency AC current, i_{hf} are shown in Fig. 3.31. The waveforms coincide with the theoretical waveforms shown in the modes of operation of the converter.

Table 3.8 Active and passive components selected for experimental hardware prototype

Component	Specification
MOSFET, S_1–S_6	FCA16N60N, 600 V, 16 A
Diode D_1, D_2	IDP15E65D2XKSA1-ND, 650 V 15 A
Input filter inductors, L_a, L_b, L_c	513-1660-ND, 200 μH 7A
Input filter capacitor, C_a, C_b, C_c	PCF1569-N, 1.2 μF 630 VDC
High frequency transformer	ETD 59, Ferrite core, 3C95
Output Inductor, L_{f1}, L_{F2}	157D-ND, 1 mH, 5.9 A
Output capacitor, C_o	Electrolytic capacitor 400 μF, 400 V
FPGA controller board	ALTERA QUARTUS II
Microcontroller board	TMDSCN28335

Fig. 3.29 C1: High frequency AC voltage, v_{hf} (200 V/div), C2: input phase-a voltage, v_{an} (200 V/div), C3: output DC voltage, V_o (20 V/div)

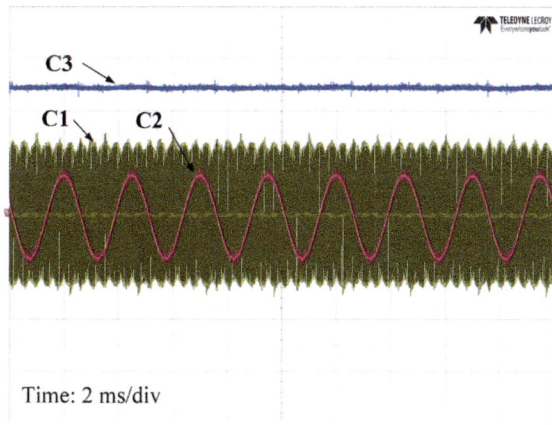

Fig. 3.30 C1: High frequency AC voltage, v_{hf} (100 V/div), C2: secondary voltage of the high frequency transformer, v_s (100 V/div)

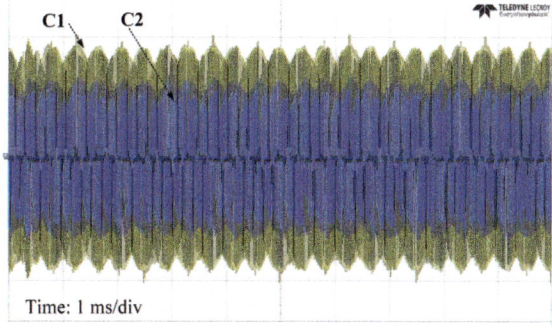

Fig. 3.31 C1: High
frequency AC voltage, v_{hf}
(100 V/div), C2: secondary
voltage of the high frequency
transformer, v_s (100 V/div),
C3: High frequency AC
current, i_{hf} (10 A/div)

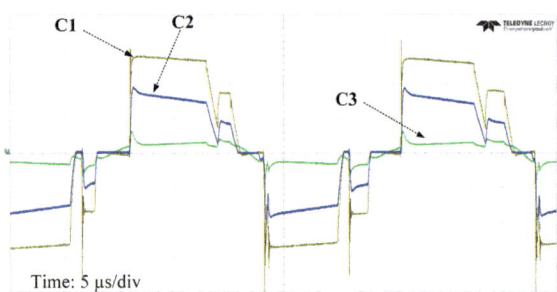

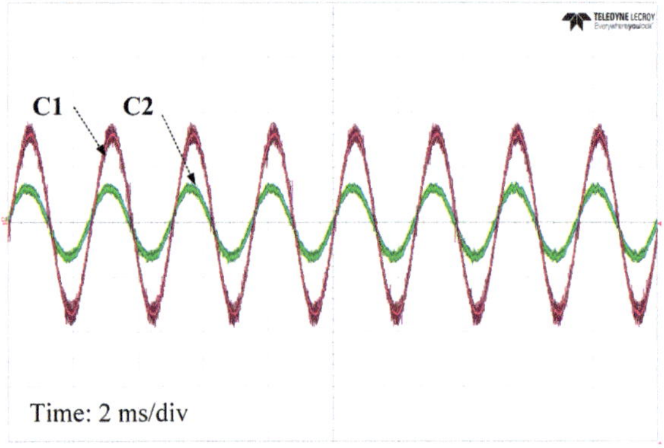

Fig. 3.32 C1: input phase-a
voltage, v_{an} (100 V/div), C2:
Input phase-a current, i_a (4
A/div)

Fig. 3.33 Input three phase
currents, C1: i_a (2 A/div),
C2: i_b (2 A/div), C3: i_c (2
A/div)

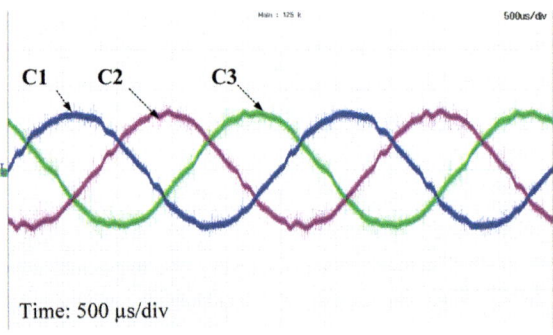

A small snubber circuit of $R_s = 56\,\Omega$ and $C_s = 0.75\,\text{nF}$ is used at the matrix output of
to reduce the high frequency voltage ringing caused by Printed Circuit Board (PCB)
and the switching device parasitics. Figure 3.32 shows the input phase-a voltage and
phase-a current at full load. The displacement power factor is found to be very close
to unity which can be further improved at increased output load. The input three
phase currents, i_a, i_b and i_c are symmetrical and balanced as shown in Fig. 3.33.
Figure 3.34 shows the output DC current, I_o and the transformer secondary current,
i_s. As discussed in the analysis of the proposed converter, the amplitude of i_s is half
of the output DC current, I_o.

Fig. 3.34 C1: transformer secondary current, i_s (4 A/div), C2: output DC current, I_o (4 A/div)

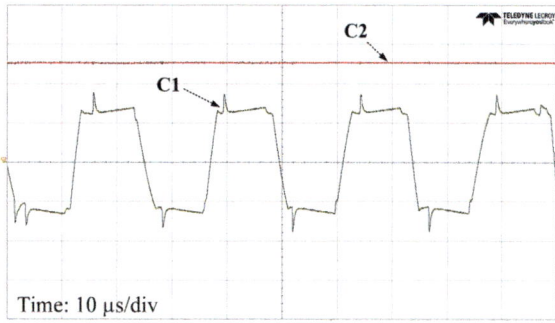

Fig. 3.35 Voltage across MOSFETs, C1: SW_{11} (100 V/div) and C2: SW_{12} (100 V/div)

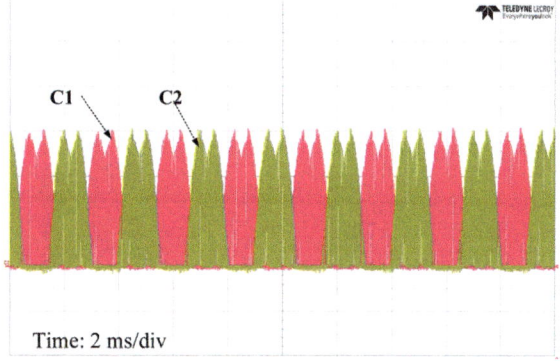

Figure 3.35 shows the voltage across the MOSFETs SW_{11} and SW_{12}. The MOS-FETs, SW_{11} and SW_{12} form the matrix switch S_1 as shown in Fig. 2.4. It is to be noted that the voltage across one of the two switches is zero during $\frac{\pi}{3}$ interval which in seconds is $\frac{\pi}{3f_i} = \frac{\pi}{1200}$ s. The zero voltage across one of the MOSFETs results in natural ZVS contributing to improved power conversion efficiency. Figure 3.36 shows the voltage across the diodes, D_1 and D_2 of the CDR circuit. During zero period, t_o, both CDR diodes, D_1 and D_2 conduct.

Figure 3.37 shows the experimental THD of phase- a current. The THD of the input current is found to be 3.06%. The THD can be further improved by increasing the value of the filter capacitor. However, the maximum value of the filter capacitors is limited by the displacement power factor. At higher load, the filter capacitors can be increased and thus, better THD is expected. For load variation from 100 to 20%, the THD has been found to vary from 3.06 to 5.8%.

Figure 3.38 shows the effect of adding capacitor in series with the primary winding of the high frequency transformer. In the absence of the series capacitor, the high frequency AC current, i_{hf} changes immediately resulting in voltage spikes due to high $\frac{di}{dt}$. Moreover, the presence of leakage inductance results in duty cycle loss as shown in Fig. 3.38a. The duty cycle loss will be more significant for higher leakage inductance as discussed in Sect. 3.5.1. Figure 3.38b shows the experimental results when series capacitor is added with the primary winding of the high frequency

Fig. 3.36 Voltage across
CDR diode, C1: D_1 (40
V/div) and C2: D_2 (40 V/div)

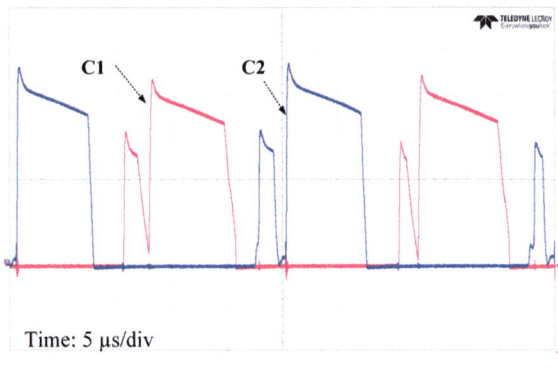

Fig. 3.37 Experimental
THD of phase-a current, i_a

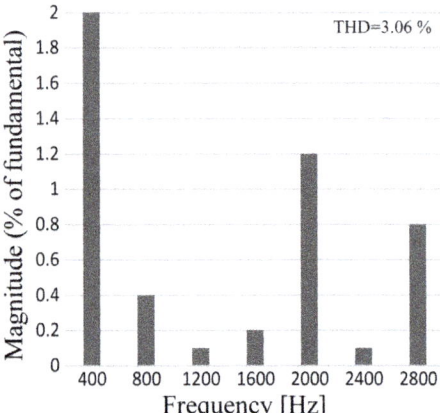

transformer. The high frequency AC current, i_{hf} changes from negative to positive
with a finite slope during zero period without any voltage spike. Moreover, duty cycle
loss is also minimized as the current, i_{hf} changes its direction and becomes positive
before the high frequency AC voltage, v_{hf} appears across the primary winding of
the high frequency transformer. A 65% reduction in duty cycle loss is demonstrated
through experimental results which can be further improved by fine tuning the series
capacitor. Thus the experimental result validates the effectiveness of the proposed
current commutation method and corroborates the theoretical analysis and simulation
result in providing soft commutation of high frequency AC current.

3.8.2   Discussion on the Power Conversion Efficiency of the Proposed Converter

Figure 3.39 shows the loss distribution of the proposed converter. The total loss of the
proposed converter can be divided into six types. The switching loss of the converter

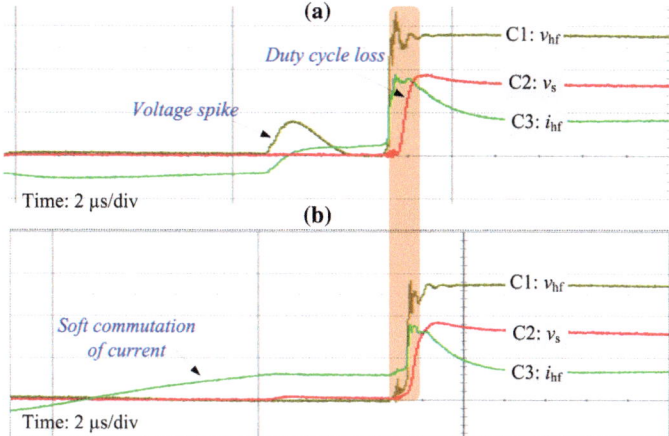

Fig. 3.38 Soft commutation of the high frequency AC current and reducing the duty loss by adding a series capacitor in presence of finite leakage inductance of the high frequency transformer, C1: v_{hf} (100 V/div), C2: v_s (100 V/div) C3: i_{hf} (4 A/div). **a** Without series capacitor. **b** With series capacitor

Fig. 3.39 Theoretical loss estimation of the proposed AC–DC converter at 500 W output power. [1]. Switching loss [2]. Conduction loss [3]. Transformer loss [4]. Input filter loss [5]. Output filter loss [6]. Snubber Loss

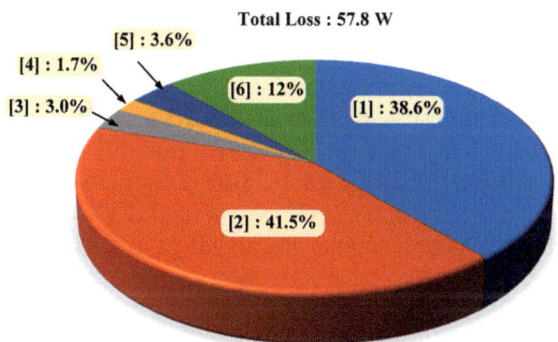

includes the switching losses of all the semiconductor devices including the matrix switches, MOSFETs in shorting legs and the CDR diodes. The switching losses of the MOSFETs include turn OFF loss and reverse recovery loss of the body diodes. As the MOSFETs in shorting legs turns ON with ZVS, therefore, the turn ON loss is considered to be zero in the loss calculation. The conduction loss includes conduction loss in all the semiconductor devices including matrix switches, MOSFETs in shorting legs and the CDR diodes. The conduction loss due to body diodes, D_{s1} and D_{s2} has been ignored in the calculation as they conduct only during the dead time, t_d. Further, the core loss and copper loss for the high frequency transformer has been calculated based on [26]. The input and output filter loss has been calculated. For simplicity, the effect of voltage and current ripple has been ignored in the loss calculation of the input and output filter. The efficiency of the proposed converter is calculated at full load (500 W). The theoretical efficiency is found to be 89.63% which is very close to experimental efficiency (87.5%). The difference between theoretical and experimental efficiency is attributed to voltage/current ripple, the ESR of the capacitor, parasitic inductances and capacitances which has been ignored in the loss calculation.

3.9 Comparison Evaluation of Isolated AC–DC Converter in Literature with the Proposed Converter

In this subsection, the benefits and contributions of the proposed matrix based AC–DC converter are discussed taking various isolated single-stage AC–DC converters proposed in literature into account. Further, the effects of mandatory DC link capacitor in conventional back to back AC–DC converter on power density, power conversion efficiency and reliability are discussed. The comparison of different design methods for single stage AC–DC converter are presented in Table 3.9. In this section, the effect of DC link capacitor on power density and reliability in a back to back AC–DC converter is investigated. First, the ripple current in DC link capacitor is derived. The ripple current heats the capacitor and maximum permitted ripple current is set by how much can be permitted and still meet the capacitor load life specification. Too much temperature at the core of the capacitors dramatically shorten the capacitor's expected life. The Equivalent Series Resistor (ESR) heats the capacitor and therefore, selection of capacitor with appropriate ESR is of paramount importance for improved life expectancy of the capacitor. Further, the design of DC link capacitor based on the current ripple and power loss is carried out. Based on the data sheet of aluminium electrolytic capacitors, the selection of capacitor is done.

3.9.1 Design of DC Link Capacitor

Generally, a back to back AC–DC converter is configured as a three phase boost rectifier in the front end followed by a DC link capacitor and a DC–DC isolated converter as shown in Fig. 3.40. A SVM based modulation scheme is used for the boost rectifier. The DC link capacitor average out the pulsating current generated at the output of the boost rectifier. In this section, first the ripple current in the DC link capacitor is calculated. Subsequently, the power loss in the DC link capacitor for different output power is estimated.

3.9.1.1 Current Ripple Calculation in DC Link Capacitor

Assuming unity power factor, the phase current is given by (3.1),

Table 3.9 Comparison of the design methods of different single stage AC–DC converter in literature

Reference [14]	1. SVM based modulation scheme
	2. Does not consider leakage inductance in design
	3. Converter efficiency is not provided
	4. Switching frequency limited to 10 kHz
Reference [27]	1. Sinusoidal Pulse Width Modulation (SPWM) based scheme
	2. No experimental results; only simulation results are provided
Reference [28]	1. Sinusoidal Pulse Width Modulation (SPWM) based scheme
	2. Uses series resonant tank with high frequency transformer
	3. Capacitive output filter
	4. Operated at 14.94 kHz switching frequency
	5. Input current THD and efficiency not mentioned
Reference [29]	1. Boost derived matrix converter
	2. Requires clamp capacitor for overvoltage protection of the matrix switch
	3. Higher voltage rating switches are required
	4. Power conversion efficiency is not provided
Reference [25]	1. A simplified modulation scheme
	2. Poor input current THD, not suitable for aircraft system
Proposed converter	1. A simplified SVM based modulation scheme
	2. Addresses the issues of current commutation and duty cycle loss
	3. Experimental efficiency 87.5% at 500 W output power
	4. THD lower than 5%
	5. 40 kHz bipolar symmetrical high frequency AC output

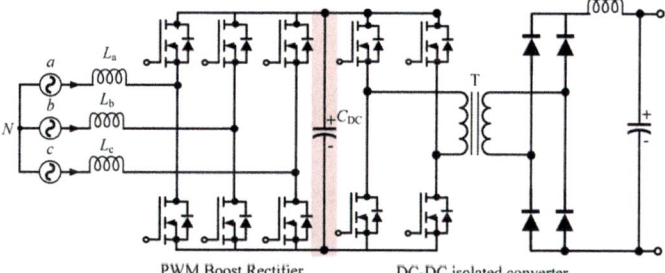

Fig. 3.40 Circuit diagram of back to back AC–DC converter. A mandatory DC link capacitor is required between the two stages of conversion

$$i_a = I_m \sin(\theta); \; i_b = I_m \sin\left(\theta - \frac{2\pi}{3}\right); \; i_c = I_m \sin\left(\theta + \frac{2\pi}{3}\right) \qquad (3.51)$$

To calculate the $I_{ripple,DC}$, the rms, $I_{o,rms}$ and avg, $I_{o,avg}$ of the pulsating output current are calculated and subsequently, $I_{ripple,DC}$ is evaluated by,

$$I_{ripple,DC} = \sqrt{I_{o,rms}^2 - I_{o,avg}^2} \qquad (3.52)$$

The timing durations for SVM for sector-1 are given by,

$$t_\alpha = mT_s \sin\left(\theta - \frac{\pi}{3}\right); \; t_\beta = mT_s \sin\left(\frac{2\pi}{3} - \theta\right) \qquad (3.53)$$

The *rms* value of the output current is calculated in two steps. In the first step, the rms value of the output current for one switching cycle, T_s is calculated. In the second step, the *rms* value for a sector ($\frac{\pi}{3}$ duration) is calculated. The rms value of current for a switching cycle in sector-1,

$$I_{o,rms,\theta} = \sqrt{\frac{i_b^2 t_\beta + (i_b + i_c)^2 t_\alpha}{T_s}} \qquad (3.54)$$

The rms value for one section duration is given using (3.54),

$$I_{o,rms} = \sqrt{\frac{3}{\pi} \int_{\frac{\pi}{3}}^{\frac{2\pi}{3}} I_{o,rms,\theta} d\theta} = I_m \sqrt{\frac{2m}{\pi}} \qquad (3.55)$$

The avg current, $I_{o,avg}$ can be similarly calculated and is derived to,

$$I_{o,avg} = \frac{3m}{4} I_m \qquad (3.56)$$

from (3.52), (3.55) and (3.56), the ripple current in DC capacitor is given as,

$$I_{ripple,DC} = \frac{3}{4} m I_m \sqrt{\frac{32}{9m\pi} - 1} \qquad (3.57)$$

for a boost AC–DC rectifier of 115 V AC rms at m = 0.8, the ripple current, $I_{ripple,DC}$ is plotted for different output power, P_o (Fig. 3.41).

3.9.1.2   Power Loss in Capacitor

The power loss in the capacitor is attributed to the ESR of capacitor. The ESR of the capacitor is dependent on the frequency and temperature. The ESR of capacitor

Fig. 3.41 Ripple current in the DC link capacitor for different output power, P_o. The input voltage, $V_{in,rms}$ is 115 V AC and the converter is operated at modulation index, $m = 0.8$

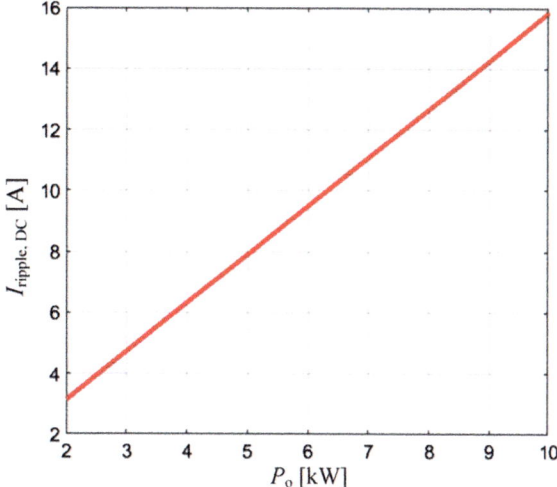

P_o [kW]

decreases at higher frequencies and higher temperatures. For a given ambient temperature, the maximum allowable temperature increase determines maximum power loss. The increased temperature of capacitors shorten the operating life of the capacitor and therefore, selection of capacitor of appropriate ESR value is very important. The power loss in capacitor is given by,

$$P_{Loss,C} = I^2_{ripple,DC} ESR \qquad (3.58)$$

For example, from [30], a 390 μF, 500 V aluminium electrolytic capacitor has ESR of 310 Ω at 20 °C for 100 Hz. At frequency more than 20 kHz, the ESR reduces to 232.5 Ω. The power loss in the capacitor for frequency more than 20 kHz at 20 °C is shown in Fig. 3.42.

3.9.2 Power Density and Reliability

The DC link capacitor is mandatory in back-to-back AC–DC converter and significantly impacts the power conversion efficiency, power density and reliability of the overall power electronic conversion system. As shown in Fig. 3.42, the loss in the capacitor can be nearly 60 W at 10 kW output power which can reduce the total efficiency of the converter by 0.6%.

The weight of the DC link capacitor is also an important consideration which becomes crucial in application such as aircraft system. It can be shown that the total weight of 390 μF, 500 V capacitor can nearly weigh 1.12 kG for 10 kW output power assuming that convection air cooling (0.006 $W/C/in^2$) is provided and allowable

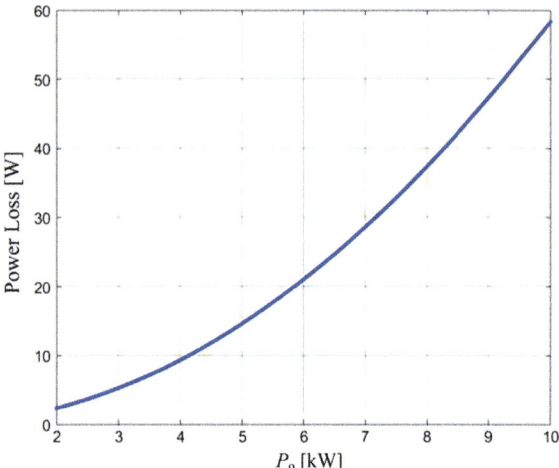

Fig. 3.42 Power loss in the DC link capacitor for different output power. The input voltage, $V_{in,rms}$ is 115 V AC and modulation index, m = 0.8

increase in capacitor temperature is $10\,°C$.[1] Further, electrolytic capacitor has limited life span. The random failure rate of the aluminum electrolytic capacitor is defined as [31],

$$\lambda = \frac{4000000N\,V_a^3\,C^{\frac{1}{2}}2^{\frac{(T_a-T_m)}{10}}}{L_b\,V_r^2} \tag{3.59}$$

where, N is number of capacitor connected in array; V_a is applied voltage in V; C is the capacitance in F; T_a is actual core temperature in degree C; T_m is maximum allowable core temperature in degree C; L_b base life in hours at T_m and V_r; V_r is rated voltage in volts. The random failure rate, λ is in FIT. The increase in temperature due to ESR loss and operating environment further shorten the life of capacitor. Overall, the presence of electrolytic capacitor reduces the reliability. It is worth noticing that from (3.57) the ripple current in DC link capacitor is independent of the switching frequency and thus, the size/volume of capacitor can be very significant compared to total size/volume of the converter at higher switching frequencies.

3.10   Benchmarking of the Proposed Converter

Table 3.10 shows the benchmarking of isolated matrix based AC–DC converter with TRU and back to back converter. TRU is basically an isolated passive rectifier whereas back to back converter is conventional two stage converter commonly used for isolated AC–DC rectification. The performance values for TRU is taken from the product datasheet [32] whereas for the proposed converter and back to back AC–DC

---

[1]ESR is reduced by 0.7 factor for high frequency and temperature increase. The size of capacitor is 35 mm (d) and 55 mm (l). The maximum allowable temperature rise is taken as $10\,°C$.

Table 3.10 Benchmarking of isolated matrix based AC–DC converter

		Type	Power density (W/Kg)	THD (%)	Efficiency (%)
1	Proposed converter	Active	**400**	**3.06**	**89.5**
2	TRU	Passive	339	<5	86
3	Back to back converter	Active	362	<5	88.44

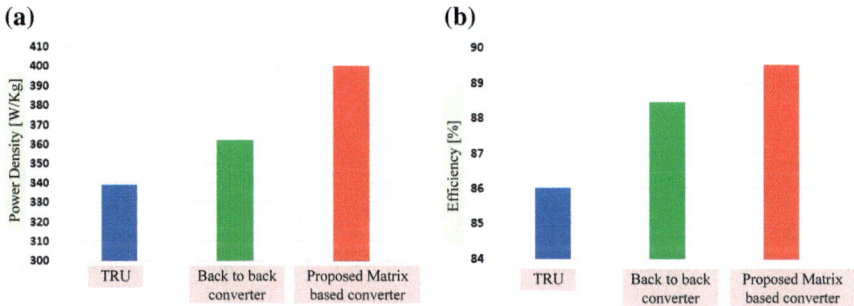

Fig. 3.43 Benchmarking for the isolated matrix based AC–DC converter. **a** Power density (W/kg). **b** Efficiency (%)

converter, the values are calculated for similar input and output specifications. The semiconductor devices are chosen similar in the both cases for comparison. Also, the switching frequency chosen is equal. It is evident from the Table 3.10 and Fig. 3.43, that the proposed converter provides highest power density and power conversion efficiency. The TRU requires line frequency multi pulse transformer both for isolation and for meeting the strict THD requirement imposed by aircraft industries. The bulky line frequency isolation transformer contributes to low power density in the TRU. Further, a back to back converter has three phase boost rectifier followed by an isolated DC–DC converter. These two stages of conversion are linked by a mandatory DC link capacitor. The boost rectifier requires three input inductors. The combined weight of the three input inductors and a mandatory DC link capacitor increases the overall weight and thus, reduces the power density. In the proposed matrix based AC–DC converter, the input inductors are very small due to the buck type configuration at the input stage. Additionally, it does not have DC link capacitor as it provides direct conversion of line frequency AC voltages to high frequency AC voltage.

The TRU shows the lowest power conversion efficiency because of bulky transformer and passive diode based rectification. The back to back converter is a two stage converter. Therefore, the net efficiency is multiplication of the efficiency of the two stages. The multiplication results in lower overall efficiency for back to back converter. For example, the calculation of overall efficiency of back to back converter shows that the first stage- three phase boost rectifier has 95.7% whereas the DC–DC converter shows 92.4% efficiency. Despite having high efficiency in both

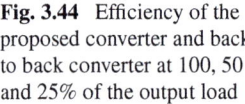

Fig. 3.44 Efficiency of the proposed converter and back to back converter at 100, 50 and 25% of the output load

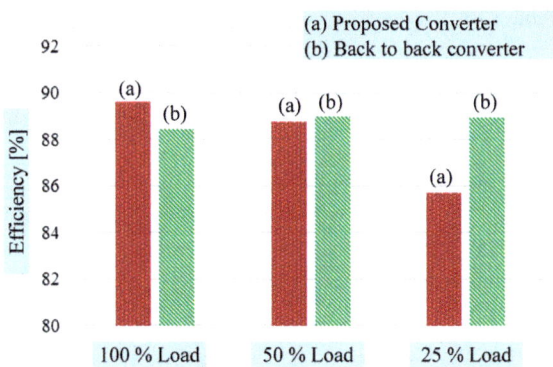

stages, the net efficiency is 88.44% which is lower than the proposed isolated matrix based AC–DC converter. Figure 3.44 shows the power conversion efficiency of the proposed converter and back to back converter for 100, 50 and 25% of the output load. The proposed converter shows higher efficiency at 100% load. However, at 50 and 25% of the load, the back to back AC–DC converter higher power conversion efficiency.

The TRU offers highest reliability due to its simplicity and no switching elements. Back to back converter is less reliable than matrix converter due to presence of mandatory bulky DC link capacitor which has limited life span and is vulnerable to temperature and pressure.

3.11 Conclusion

In this chapter, a single-stage matrix based isolated three phase AC–DC converter suitable for aircraft systems has been presented. The steady-state operation, analysis, and design are illustrated. Simulation and experimental results clearly confirm and demonstrate the claimed soft commutation of high frequency AC current and minimization of duty cycle loss along with superior input power quality. The use of matrix topology allows direct conversion of the input three phase AC voltages into high frequency AC voltage and thus, provides single-stage conversion without any intermediate bulky DC link capacitor. It has been theoretically calculated that the weight of DC link capacitor can be up to 1.12 kg for 10 kW output power. In the proposed switching scheme, the body diodes of the matrix switches do not conduct and thus, the power loss due to body diodes are eliminated. Moreover, one of the two MOSFETs of the matrix switch achieves natural ZVS and consequently, the switching loss of the proposed converter is minimized. The proposed converter requires an additional shorting leg to circulate the leakage energy of the high frequency transformer. It has been demonstrated that 65% reduction in duty cycle loss is achieved through the proposed current commutation scheme. However, the two MOSFETs of the shorting leg turn ON with ZVS and only conduct current during the zero period and dead time, and therefore, do not impact the power conversion efficiency significantly. The experimental power conversion efficiency of the proposed converter is found to be 87.5% at 500 W output power. The two output inductors in the CDR circuits act very similar to interleaved boost inductor and therefore, reduces the peak-to-peak magnitude of output current ripple and doubles the output current ripple frequency contributing to reduced size of the output filter capacitor. Since the high frequency AC current achieves soft commutation, the EMI due to high $\left[\frac{di}{dt}\right]$ is eliminated. A superior input current THD of 3.06% is demonstrated through experimental results. Additionally, single control for a matrix switch simplifies the digital implementation of switching scheme at very high switching frequency. The switching waveforms of the proposed converter can be further improved by proper layout design of the PCB. The single-stage three phase AC–DC conversion with galvanic isolation makes the converter promising for the applications where high power density, high power conversion efficiency and high input power quality are required. These merits make the proposed converter suitable for aircraft system, telecommunication, micro-grid and energy storage.

References

1. J. Rosero, J. Ortega, E. Aldabas, L. Romeral, Moving towards a more electric aircraft. IEEE Aerosp. Electron. Syst. Mag. **22**, 3–9 (2007)
2. P. Wheeler, S. Bozhko, The more electric aircraft: technology and challenges. IEEE Electrif. Mag. **2**, 6–12 (2014)
3. B. Sarlioglu, C. Morris, More electric aircraft: review, challenges, and opportunities for commercial transport aircraft. IEEE Trans. Transp. Electrif. **1**, 54–64 (2015)
4. Y. Deng, S. Foo, I. Bhattacharya, Regenerative electric power for more electric aircraft, in *IEEE SOUTHEASTCON 2014* (2014), pp. 1–5
5. B. Sarlioglu, Advances in ac-dc power conversion topologies for more electric aircraft, in *2012 IEEE Transportation Electrification Conference and Expo (ITEC)* (2012), pp. 1–6
6. R. Jones, The more electric aircraft: the past and the future?, in *IEE Colloquium on Electrical Machines and Systems for the More Electric Aircraft (Ref. No. 1999/180)* (1999), pp. 1/1–1/4
7. R. Naayagi, A review of more electric aircraft technology, in *2013 International Conference on Energy Efficient Technologies for Sustainability (ICEETS)* (2013), pp. 750–753
8. J. Vieira, J. Oliver, P. Alou, J. Cobos, Power converter topologies for a high performance transformer rectifier unit in aircraft applications, in *2014 11th IEEE/IAS International Conference on Industry Applications (INDUSCON)* (2014), pp. 1–8
9. J. Lee, Aircraft transformer-rectifier units. Stud. Q. J. **42**, 69–71 (1972)
10. J. Kolar, T. Friedli, The essence of three-phase pfc rectifier systems-part i. IEEE Trans. Power Electron. **28**, 176–198 (2013)
11. T. Friedli, M. Hartmann, J. Kolar, The essence of three-phase pfc rectifier systems-part ii. IEEE Trans. Power Electron. **29**, 543–560 (2014)
12. F. Xu, B. Guo, L. Tolbert, F. Wang, B. Blalock, An all-sic three-phase buck rectifier for high-efficiency data center power supplies. IEEE Trans. Ind. Appl. **49**, 2662–2673 (2013)
13. Y. Zhang, L. Jin, Y. Jing, Z. Zhao, T. Lu, Three-level pwm rectifier based high efficiency batteries charger for ev, in *2013 IEEE Vehicle Power and Propulsion Conference (VPPC)* (2013), pp. 1–4

14. S. Ratanapanachote, H. J. Cha, P. Enjeti, A digitally controlled switch mode power supply based on matrix converter, in *2004 IEEE 35th Annual Power Electronics Specialists Conference, 2004. PESC 04*, vol. 3 (2004), pp. 2237–2243

15. J. Sandoval, S. Essakiappan, P. Enjeti, A bidirectional series resonant matrix converter topology for electric vehicle dc fast charging, in *2015 IEEE Applied Power Electronics Conference and Exposition (APEC)* (2015), pp. 3109–3116

16. Y. Wang, L. Yang, C. Wang, Z. Meng, Three phase high step-up single-stage flyback converter with no electrolytic capacitor, in *2014 International Electronics and Application Conference and Exposition (PEAC)* (2014), pp. 1207–1211

17. B. Tamyurek, D. Torrey, A three-phase unity power factor single-stage ac-dc converter based on an interleaved flyback topology. IEEE Trans. Power Electron. **26**, 308–318 (2011)

18. T. Zhao, J. Su, D. Xu, M. Mao, Commutation compensation for matrix based rectifiers due to leakage inductances of isolation transformer, in *2015 9th International Conference on Power Electronics and ECCE Asia (ICPE-ECCE Asia)* (2015), pp. 1803–1808

19. H. Keyhani, H. Toliyat, Isolated zvs high-frequency-link ac-ac converter with a reduced switch count. IEEE Trans. Power Electron. **29**, 4156–4166 (2014)

20. H. Keyhani, H. Toliyat, W. Alexander, A single-stage multi-string quasi-resonant inverter for grid-tied photovoltaic systems, in *2013 IEEE Energy Conversion Congress and Exposition (ECCE)* (2013), pp. 1925–1932

21. H. Keyhani, H. Toliyat, M. Harfman-Todorovic, R. Lai, R. Datta, An isolated resonant ac-link three-phase ac-ac converter using a single hf transformer. IEEE Trans. Ind. Electron. **61**, 5174–5183 (2014)

22. H. Keyhani, H. Toliyat, M. Todorovic, R. Lai, R. Datta, Step-up down three-phase resonant high-frequency ac-link inverters. IET Power Electron. **7**, 1246–1255 (2014)

23. H. Keyhani, M. Johnson, H. Toliyat, A soft-switched highly reliable grid-tied inverter for pv applications, in *2014 Twenty-Ninth Annual IEEE Applied Power Electronics Conference and Exposition (APEC)* (2014), pp. 1725–1732

24. H. Keyhani, H. Toliyat, Single-stage multistring pv inverter with an isolated high-frequency link and soft-switching operation. IEEE Trans. Power Electron. **29**, 3919–3929 (2014)

25. N.X. Bac, D. Vilathgamuwa, U. Madawala, A sic-based matrix converter topology for inductive power transfer system. IEEE Trans. Power Electron. **29**, 4029–4038 (2014)

26. C.H. Yang, T.J. Liang, K.H. Chen, J.S. Li, J.S. Lee, Loss analysis of half-bridge llc resonant converter, in *2013 1st International Future Energy Electronics Conference (IFEEC)* (2013), pp. 155–160

27. H. Krishnamoorthy, P. Garg, P. Enjeti, A matrix converter-based topology for high power electric vehicle battery charging and v2g application, in *IECON 2012 - 38th Annual Conference on IEEE Industrial Electronics Society* (2012), pp. 2866–2871

28. J.J. Sandoval, S. Essakiappan, P. Enjeti, A bidirectional series resonant matrix converter topology for electric vehicle dc fast charging, in *2015 IEEE Applied Power Electronics Conference and Exposition (APEC)* (2015), pp. 3109–3116

29. C. Li, Y. Zhong, D. Xu, Soft-switching three-phase matrix based isolated ac-dc converter for dc distribution system, in *2015 IEEE Energy Conversion Congress and Exposition (ECCE)* (2015), pp. 6755–6761

30. Aluminum electrolytic capacitors, https://media.digikey.com/pdf/Data%20Sheets/Epcos%20PDFs/B43510_B43520_Rev_Jun_2015.pdf. [Online]

31. Aluminum Electrolytic Capacitor Application Guide, http://www.cde.com/resources/catalogs/AEappGUIDE.pdf. [Online]

32. Transformer rectifier unit, http://www.powercontrolsystemsgroup.com/pdf/b804.pdf. [Online]

Chapter 4
A New Matrix Based Non-isolated Three Phase Buck-Boost Rectifier

4.1 Introduction

In this chapter, a new matrix based non-isolated three phase buck-boost rectifier is proposed for aircraft application. Similar to the converter topologies presented in Chaps. 2 and 3, this topology uses matrix converter topology for three phase line frequency AC voltage to single phase high frequency AC voltage conversion. Being a buck-boost converter, the proposed converter provides wide range of the output voltage. The chapter is divided into seven sections. In Sect. 4.2, the brief review of the power converter for MEA is discussed. The new contributions of the chapter are highlighted in Sect. 4.3. Section 4.4 presents the topology and operation of the converter in details. In Sect. 4.5, the comprehensive steady state analysis and design of the converter are discussed. Comparative evaluation of the proposed converter with the boost-buck type of rectifier is discussed in Sect. 4.6. In Sect. 4.7, a scale down hardware prototype of the proposed converter is built and experimental test results are demonstrated to validate the theoretical claims. Section 4.8 provides the conclusion.

4.2 A Brief Review of Three Phase Buck-Boost AC–DC Converter

In modern aircrafts, the 230 V rms AC bus is used instead of 115 V rms AC. The increased voltage reduces the conductor weight almost by 35% [1, 2]. However, with increase in the input voltage, the aircraft needs to push the output voltage from 270 V DC to 540 V DC.

In case of 540 VDC output, two electrical architectures, $\pm/0$ 270 VDC and ± 270 VDC are possible which are shown in Fig. 4.1a, b. In case of electrical architecture shown in Fig. 4.1a, two different types of load, one for positive and other for negative

© Springer Nature Singapore Pte Ltd. 2018
A. K. Singh, *Analysis and Design of Power Converter Topologies for Application in Future More Electric Aircraft*, Springer Theses, https://doi.org/10.1007/978-981-10-8213-9_4

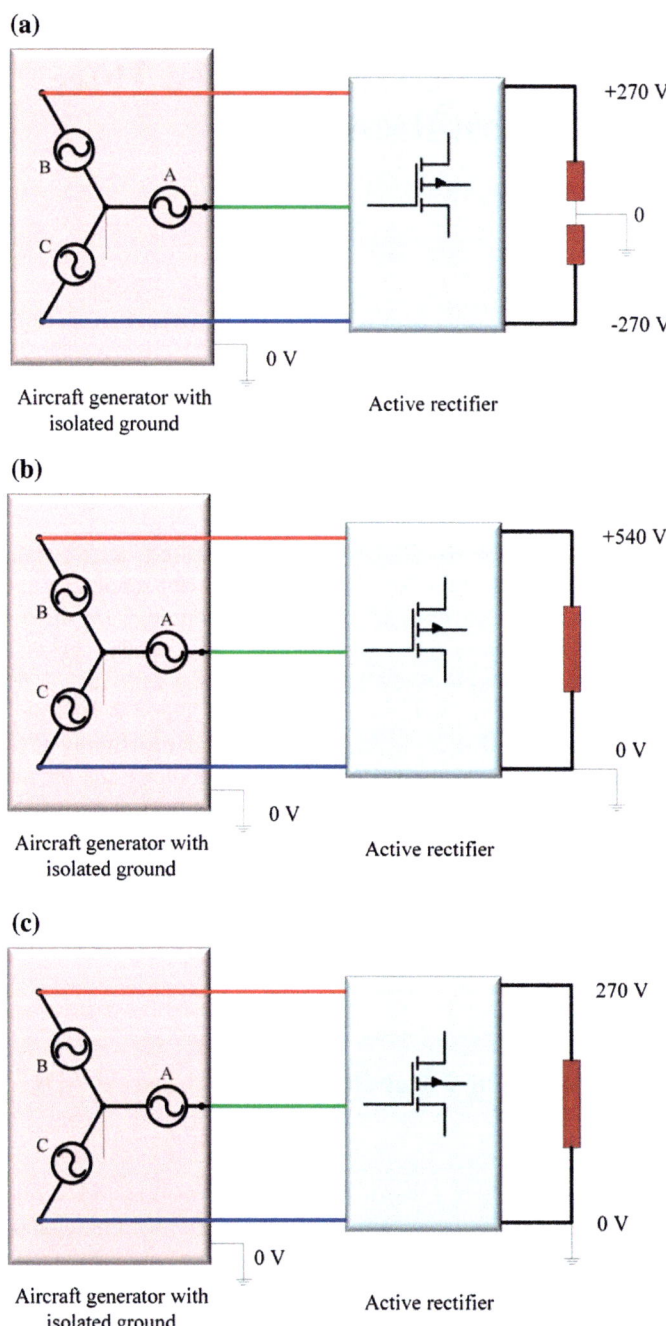

Fig. 4.1 Different possible electrical architecture for 230 V AC to 270 V DC conversion. **a** ±/0 270 V DC. **b** ±270 V DC. **c** 0/270 V DC

is required. For the electrical architecture shown in Fig. 4.1b, the loads of aircrafts are to be modified for 540 V DC which requires increased insulation and higher voltage rated devices.

The grounded 270 V DC based architecture is shown in Fig. 4.1c. In [2], the cable weights of different DC and AC voltage possibilities are compared and it has been found that the grounded 270 V DC architecture with decentralized 28 V DC generation is the best choice for improved power density compared to $\pm/0\,270$ V DC and ± 270 V DC architecture. Moreover, the boost type of rectifiers can be used for rectifying ± 270 VDC which has more control complexity, higher weight and less reliability compared to a buck type of rectifier [3]. The balancing of the two output capacitor voltages adds further complexity in ± 270 VDC architecture [4].

The rectification of three phase 230 V rms AC to 270 V DC requires a voltage gain, $G_v \left[\frac{V_o}{V_{in,rms}} \right]$ of 1.17. For passive rectifiers, the change in the winding ratio of the auto-transformer is enough to get desired voltage gain. However, in case of non-isolated PWM active rectifiers, both the PWM boost rectifiers and the PWM buck rectifiers are not suitable for such voltage gain. As shown in Fig. 4.2, the boost rectifier has a lower limit over minimum output voltage $[2\sqrt{2}\,V_{in,rms}]$ and therefore, the voltage gain, G_v of 1.17 cannot be achieved. For the buck type of rectifiers, the voltage gain, G_v of 1.17 is possible. However, the buck rectifier has to be operated at very low modulation index ($m = 0.55$) for required voltage gain which is not preferable due to the poor input power quality and increased switch rms current.

The buck-boost or boost-buck topologies are used for extended voltage gain [5–7]. In [5, 6] buck-boost and boost-buck rectifiers are compared and it has been

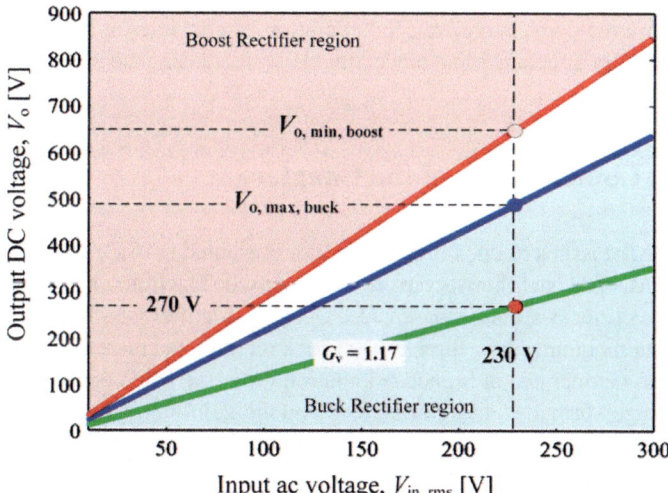

Fig. 4.2 Voltage gain range of the conventional three phase buck, boost converter and matrix based buck-boost converter. Voltage gain, G_V is the ratio of output voltage, V_o and input rms voltage, $V_{in,rms}$. m is the modulation index

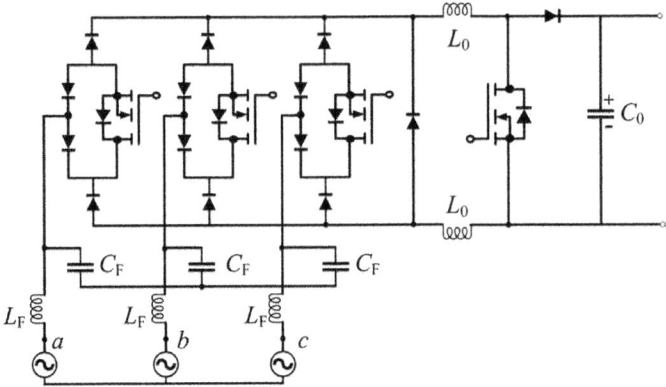

Fig. 4.3 Topology of buck-boost converter presented in [6]

demonstrated that the buck-boost converter has higher power density and reduced control complexity than boost-buck converter.

Figure 4.3 shows the buck-boost converter presented in [6]. It consists of a three switch buck rectifier followed by a boost stage. This topology requires less number of switches and provides higher power density than a boost-buck rectifier for the same input–output specifications. However, for the voltage gain, $G_v = 1.17$, this converter has following limitations.

- As the front end of the converter is a three switch buck type rectifier, the voltage gain, $G_v = 1.17$ requires that converter should be operated at the modulation index, $m \leq 0.55$ which increases switch rms currents contributing to reduced power conversion efficiency. Moreover, the operation of converter at lower modulation index adversely effect the input power quality by increasing the input current THD.

4.3   New Contributions of the Chapter

To overcome the issues of buck boost converter presented in [6], a new matrix based three phase AC–DC buck-boost converter is proposed. The front end of the proposed buck-boost rectifier is a matrix based AC–DC buck rectifier which provides half of the voltage gain obtained by a three-switch buck rectifier and therefore, the proposed converter can be operated at higher modulation index, m. The combination of buck and boost stages provides a large range of voltage gain with superior input power quality and power conversion efficiency. Even though the proposed topology requires significantly larger number of switches compared to buck-boost converter proposed in [6], it has significantly lower number of diodes. Moreover, a SVM based modulation scheme is presented for the proposed buck-boost rectifier which avoids body diode conduction and thus promises lower switch conduction loss. The presented modulation scheme is very suitable for SiC MOSFETs which exhibit very low on resistance but high forward voltage drop in the body diode. The low on resistance of SiC MOSFET results in lower switch conduction loss. The important contributions of the chapter are summarized as follows:

- proposes a new matrix based three phase buck-boost converter topology suitable for 230 V AC to 270 V DC conversion;
- discusses the principles of the operation of the proposed converter in details followed by comprehensive analysis and design;
- similar to Chaps. 2 and 3, SVM based modulation scheme at high switching frequency is implemented for the matrix converter. Moreover the benefits of the proposed converter over three phase boost-buck converter [5, 6] is discussed; and
- the operation and performance of the proposed converter prototype are validated using a scale down hardware prototype.

4.4 Topology and the Principles of Operation

In this section, the topology of the converter is described. Subsequently, the principles of the operation of the proposed converter are discussed in details.

4.4.1 Topology Description of the Proposed Converter

Figure 4.4 shows the circuit of the proposed matrix based three phase buck-boost converter. The three phase input voltages, v_{an}, v_{bn} and v_{cn} are filtered by L-C input

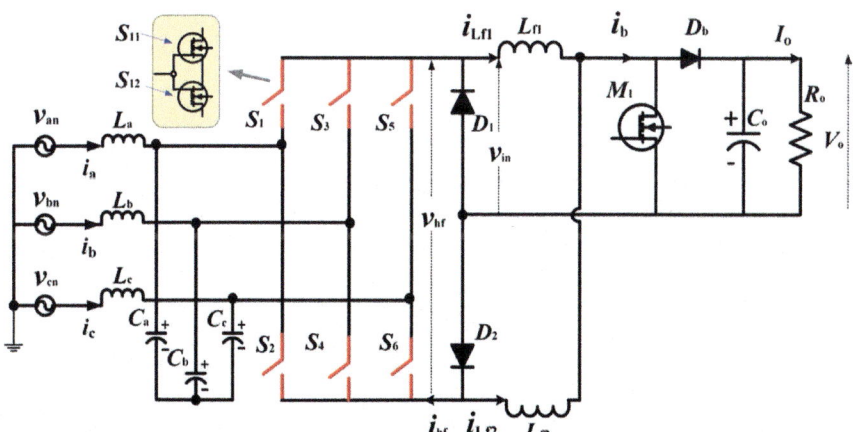

Fig. 4.4 Topology of the proposed matrix based buck-boost converter

filter formed by the three inductors, L_a, L_b and L_c and the capacitors, C_a, C_b and C_c. The filtered three phase voltages are fed to the matrix (3×1) topology formed by six bidirectional switches, S_1–S_6. Each of the matrix switch is formed by connecting two back to back MOSFETs. For a given gating signal, switches are either OFF or ON. No current flows through the switches during OFF states. Similarly, during ON conditions of the switch, current flows through the MOSFET channel and therefore, switch diodes never carry current resulting in lower switch conduction loss. The high frequency ac voltage, v_{hf} is processed using the CDR circuit formed by the two diodes, D_1 and D_2 and the two filter inductors, L_{f1} and L_{f2}. Subsequently, a boost stage is added with the switch, M_1, the diode, D_b, the output filter capacitor, C_o and resistor, R_o.

4.4.2 Principles of Operation

The principles of the operation of the proposed converter are divided into two parts. The first part discusses the operation of matrix converter whereas in second part the operation of the boost stage is discussed.

4.4.2.1 Operation of the Matrix Converter

The matrix converter is used to directly convert the line frequency AC voltages into high frequency AC voltage. The high frequency AC voltage is then processed using a CDR circuit to rectify the AC voltage. A SVM based modulation scheme is used for the matrix converter to achieve high input power quality with reduced switch conduction losses. The three phase AC voltages are given by,

$$v_{an} = V_m \sin \theta; v_{bn} = V_m \sin \left(\theta - \frac{2\pi}{3} \right);$$

$$v_{cn} = V_m \sin \left(\theta + \frac{2\pi}{3} \right) \tag{4.1}$$

where, V_m is the peak of the input voltages and θ is in radians. To implement the SVM scheme, the input three phase voltages are divided into six similar sectors as shown in Fig. 4.5. In each sector, the vector I_{ref} is synthesized by using two adjacent vectors. Based on the magnitude and angle of I_{ref}, the time duration t_α and t_β are calculated. For a given sector, the appropriate switches as shown in Fig. 4.7c are operated. The values of t_α and t_β vary during the sector and pulse width modulated currents are formed which are subsequently filtered using LC filter to generate high quality (low THD$_i$) currents at the input. For Sector-1, the current reference, I_{ref} is given using Fig. 4.5,

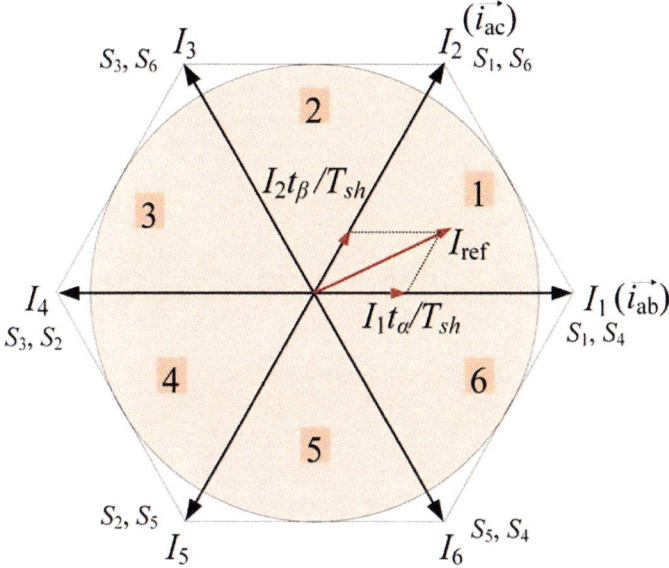

Fig. 4.5 Space Vector Modulation scheme for the presented modulation scheme. $T_{sh} = \frac{T_s}{2}$

$$I_{ref} = i_{ab}t_\alpha + i_{ac}t_\beta \tag{4.2}$$

Lets assume that V_{dc}^* voltage across the matrix output. When current i_{ab} flows, voltage v_{ab} ($v_{an} - v_{bn}$) appears across the matrix output. Similarly, for current i_{ac}, voltage v_{ac} ($v_{an} - v_{cn}$) appears across the matrix output. From Fig. 4.6, following equation can be derived,

$$V_{dc}^* T_{sh} = v_{ab}t_\alpha + v_{ac}t_\beta \tag{4.3}$$

from geometry,

$$cos\phi = \frac{v_{ab}t_\alpha + V_{dc}^* sin\phi tan30°}{V_{dc}^* T_{sh}} \tag{4.4}$$

Sector-1 ranges from $\theta = \frac{\pi}{3}$ to $\theta = \frac{2\pi}{3}$. The relationship between θ and ϕ can be given as,

$$\theta = \frac{\pi}{3} + \phi \tag{4.5}$$

From (4.4) and (4.5), t_α is calculated as,

$$t_\alpha = m\frac{T_s}{2} \sin\left(\frac{2\pi}{3} - \theta\right) \tag{4.6}$$

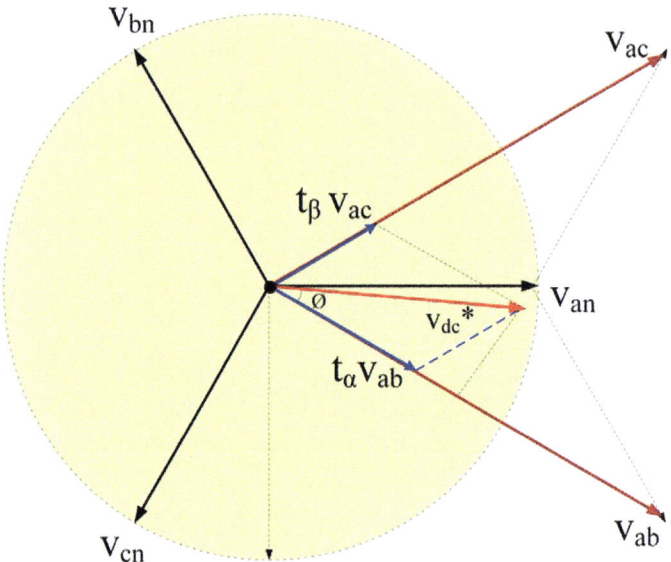

Fig. 4.6 Calculation of t_α and t_β using basic geometry

where, m is

$$m = \frac{2V_{dc}^*}{3V_m} \tag{4.7}$$

Similarly, t_β is computed as,

$$t_\beta = m\frac{T_s}{2} \sin\left(\theta - \frac{\pi}{3}\right) \tag{4.8}$$

Figure 4.7 shows the operation of the proposed matrix converter. The duration t_o is defined by,

$$t_o = \frac{T_s}{2} - t_\alpha - t_\beta \tag{4.9}$$

The calculation of t_α and t_β is shown in Table 4.1 for the different sectors. During each sector, the time duration, t_o is calculated using (4.9). As shown in Fig. 4.7a, the timing durations are given in the form of x_1, x_2 and x_3 which is subsequently, compared with saw tooth signal. The frequency of the sawtooth signal is twice the frequency of frequency of high frequency ac output of the matrix topology. An inversion signal, U as shown in Fig. 4.7b is used to generate symmetrical bipolar high frequency voltage. The states of the switches of the matrix converter are shown in Fig. 4.7c. The interval, $(t_0, t_1]$ is free wheeling interval in which both the diodes, D_1 and D_2 conducts and carries half of the output current resulting in $v_{hf} = 0$ and $i_{hf} = 0$. During duration $(t_1, t_2]$, the switches S_1 and S_4 are ON giving rise to v_{ab} ($v_{an} - v_{bn}$) across the matrix output. For duration, $(t_2, t_3]$, the matrix switches S_4 is

Fig. 4.7 Operation of the proposed matrix converter during sector-1. $T_s = 2 (t_\alpha + t_\beta + t_o)$. **a** Sawtooth signal. **b** Inversion signal (U). **c** Switching states of the matrix switches. **d** High frequency ac voltage, v_{hf}

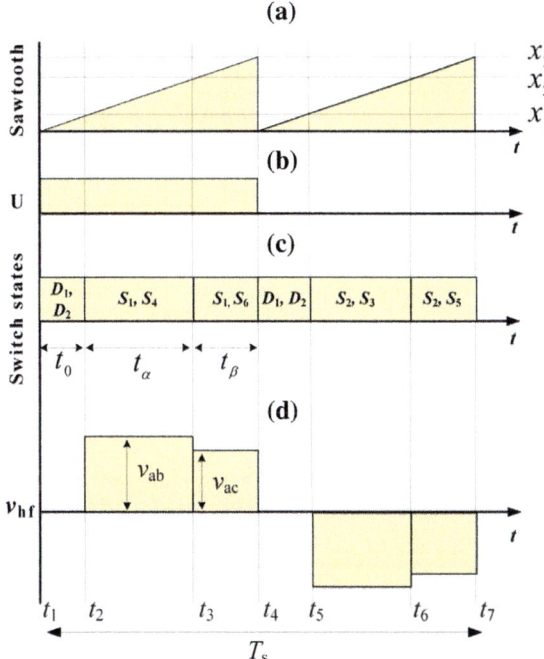

Table 4.1 Timing calculation for the different sector

Sector	t_α	t_β
1- ($\frac{\pi}{3} < \theta \le \frac{2\pi}{3}$)	$\frac{mT_s}{2} \sin\left(\theta - \frac{\pi}{3}\right)$	$\frac{mT_s}{2} \sin\left(\frac{2\pi}{3} - \theta\right)$
2- ($\frac{2\pi}{3} < \theta \le \pi$)	$\frac{mT_s}{2} \sin\left(\theta - \frac{2\pi}{3}\right)$	$\frac{mT_s}{2} \sin\left(\pi - \theta\right)$
3- ($\pi < \theta \le \frac{4\pi}{3}$)	$\frac{mT_s}{2} \sin\left(\theta - \pi\right)$	$\frac{mT_s}{2} \sin\left(\frac{4\pi}{3} - \theta\right)$
4- ($\frac{4\pi}{3} < \theta \le \frac{5\pi}{3}$)	$\frac{mT_s}{2} \sin\left(\theta - \frac{4\pi}{3}\right)$	$\frac{mT_s}{2} \sin\left(\frac{5\pi}{3} - \theta\right)$
5- ($\frac{5\pi}{3} < \theta \le 2\pi$)	$\frac{mT_s}{2} \sin\left(\theta - \frac{5\pi}{3}\right)$	$\frac{mT_s}{2} \sin\left(2\pi - \theta\right)$
6- ($0 < \theta \le \frac{\pi}{3}$)	$\frac{mT_s}{2} \sin(\theta)$	$\frac{mT_s}{2} \sin\left(\frac{\pi}{3} - \theta\right)$

turned OFF and the switch S_6 is turned ON which gives rise to v_{ac} ($v_{an} - v_{cn}$) across the matrix output. The interval $(t_3, t_4]$ is exactly similar to The interval, $(t_0, t_1]$ Both diodes D_1 and D_2 provides free-wheeling path for the inductor currents, i_{Lf1} and i_{Lf2}. To get symmetrical bipolar high frequency ac voltage, the switching operation of the matrix converter is inverted in (t_4, t_6) interval. During the time interval $(t_4, t_5]$, the switches, S_2 and S_3 are turned ON giving rise to $-v_{ab}$ across the matrix output for t_α duration. Similarly, during the time interval $(t_5, t_6]$, the switches, S_2 and S_5 are

Fig. 4.8 Simplified power
circuit for the boost stage of
the proposed converter

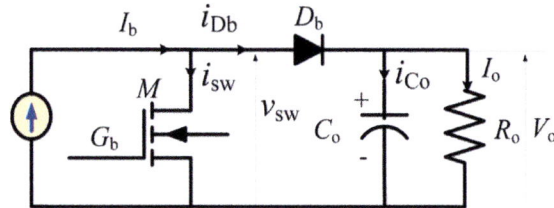

turned ON giving rise to $-v_{ac}$ across the matrix output for t_β duration. The operation
from t_0 to t_6 completes the operation of the converter for one switching period, T_s.

4.4.2.2   Operation of the Boost Converter

The equivalent circuit of the boost stage of the proposed converter is illustrated in
Fig. 4.8. The operation of the boost converter is shown in Fig. 4.9. Assuming the
current in inductors, L_{f1} and L_{f2} to be constant, the operation of the converter is
shown for a switching period, T_{sb}. The input current, i_b which is sum of the inductor
currents, i_{Lf1} and i_{Lf2} is switched by the MOSFET, M_1 with switching frequency,
$f_{sb} = \frac{1}{T_{sb}}$. During ON period of the switch M_1, the inductors, L_{f1} and L_{f2} charges
which is subsequently, discharged to load through the diode, D_b during OFF duration
resulting in the boost operation of the converter. The voltage across the switch, v_{sw},
the current, i_b, the current through the diode, D_b and the output capacitor current,
i_{Co} are shown in Fig. 4.9.

4.5   Steady State Analysis and Design of the Proposed Converter

In this section, the steady state analysis of the proposed converter is carried out
and voltage gain of the converter is derived. Subsequently, the design equations are
derived for the proposed ac–dc converter.

4.5.1   Derivation of the Different Parameters of the Converter

The proposed converter converts three phase line frequency voltages into high frequency voltage which is further rectified and boosted using an integrated boost converter. The voltage gain of the converter is defined as the ratio of output DC voltage
and input rms AC voltage. Following are the assumption taken into consideration for
the derivation of the voltage gain for the proposed converter.

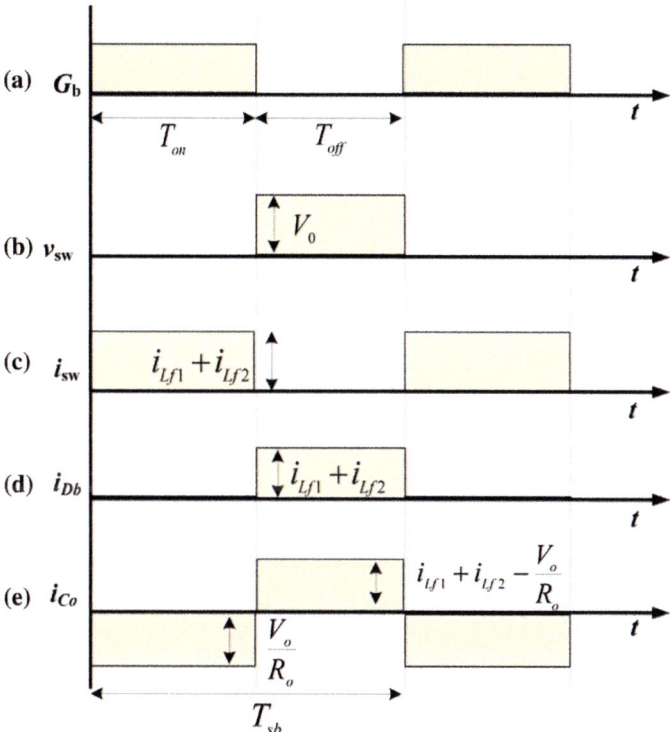

Fig. 4.9 a Gating signal for the MOSFET, M. **b** Voltage across the boost MOSFET, M (v_{sw}). **c** Current through the boost MOSFET, M (i_{sw}). **d** Current through the diode, D_b (i_{Db}). **e** Current through the capacitor, C_o (i_{Co})

- All the three input phases are balance and the input phase voltages and currents are in phase i.e. unity power factor;
- the currents in the inductors, i_{Lf1} and i_{Lf2} are ripple free and constant and the two inductors, L_{f1} and L_{f2} are identical;
- the output voltage, V_o is ripple free and constant;
- all the passive and active components are assumed to be ideal.

The current, i_b as shown in Fig. 4.8 is the sum of the currents, i_{Lf1} and i_{Lf2}.

$$I_b = i_{Lf1} + i_{Lf2} \tag{4.10}$$

In steady state, the total change in charge through the capacitor during one switching cycle should be zero which when applied on the output capacitor, C_o results in,

$$\left(-\frac{V_o}{R_o}\right)T_{on} + \left(I_b - \frac{V_o}{R_o}\right)T_{off} = 0 \tag{4.11}$$

Table 4.2 Voltage across inductor, L_{f1}

Duration	Voltage across inductor, L_{f1}
t_0–t_1	$-v_{sw}(t)$
t_1–t_2	$v_{ab}(t) - v_{sw}(t)$
t_2–t_3	$v_{ac}(t) - v_{sw}(t)$
t_3–t_4	$-v_{sw}(t)$
t_4–t_5	$-v_{sw}(t)$
t_5–t_6	$-v_{sw}(t)$

Simplifying (4.11) results in,

$$I_b = \frac{I_o}{1 - D} \quad \text{where, } D = \frac{T_{on}}{T_{on} + T_{off}} \tag{4.12}$$

where, I_o is the output current and D is the duty cycle of the boost stage. The current flowing through each of the inductors, L_{f1} and L_{f2} is $\frac{I_b}{2}$. The current $\frac{I_b}{2}$ flows through the matrix switches and essentially synthesizes the input phase current. To derive the voltage gain of the proposed converter, volt-time balance across one of the CDR inductors is carried out. For L_{f1} the voltage across inductor for one switching cycle, T_s is given as (Table 4.2).

The volt-time balance across inductor, L_{f1} results in,

$$-v_{sw}(t)t_o + (v_{ab}(t) - v_{sw}(t))t_\alpha + (v_{ac}(t)$$
$$- v_{sw}(t))t_\beta - v_{sw}(t)\left(\frac{T_s}{2}\right) = 0 \tag{4.13}$$

if D is the duty cycle of the boost converter, for $f_{sb} > f_s$, (4.13) can be simplified as,

$$\frac{3}{2}m V_m \frac{T_s}{2} = V_o(1 - D)T_s \tag{4.14}$$

From (4.14), the voltage gain of the proposed converter is derived as,

$$G_v = \frac{V_o}{V_{in,rms}} = \frac{3}{2\sqrt{2}} \frac{m}{(1 - D)} \tag{4.15}$$

where, V_m is the peak of the input phase voltage. As shown in (4.15), the output voltage, V_o depends on the modulation index, m of the matrix converter and the duty cycle, D of the boost stage. By controlling these two parameters independently, the desired output voltage can be obtained. Figure 4.10 shows the voltage gain of the proposed converter at different modulation index, m and duty, D. The yellow dotted line shows the values of m and D for which the proposed converter provides a voltage gain, G_v of 1.17 which is desired for 230 V AC to 270 V DC conversion.

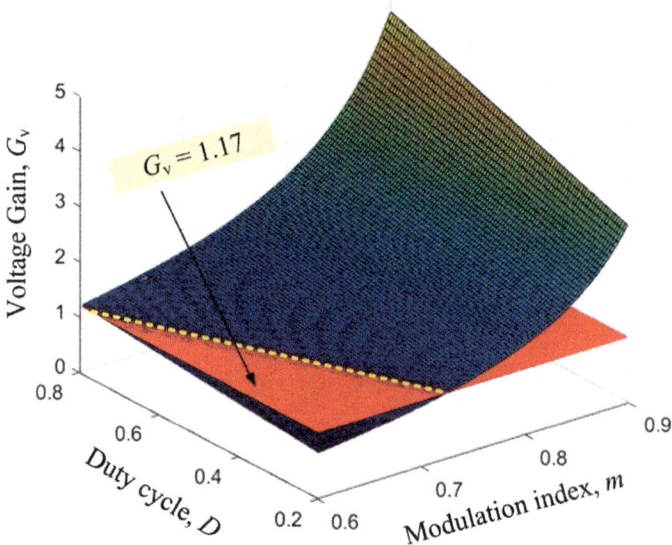

Fig. 4.10 Voltage Gain, G_v of the proposed converter at different modulation index, m and duty, D. Modulation index, m range is [0.6–0.9] and duty cycle, D range is [0.2–0.8]

Table 4.3 Voltage stress across semiconductor devices

Components	Maximum voltage
Matrix switches, S_1–S_6	$\sqrt{3}V_m$
Boost stage MOSFET, M_1	V_o
CDR diodes (D_1 and D_2)	$\sqrt{3}V_m$
Boost diode D_b	V_o

Table 4.4 Current stress of the semiconductor devices

Components	rms	avg
Matrix witches, S_1–S_6	$\frac{I_o}{2(1-D)}\sqrt{\frac{m}{\pi}}$	$\frac{I_o}{2(1-D)}\frac{m}{\pi}$
Boost stage MOSFET, M_1	$I_o\frac{\sqrt{D}}{1-D}$	$I_o\frac{D}{1-D}$
CDR diodes (D_1 and D_2)	$\frac{I_o}{\sqrt{2}(1-D)}$	$\frac{I_o}{2(1-D)}$
Boost diode D_b	$\frac{I_o}{\sqrt{1-D}}$	I_o

4.5.2 Design Equations for the Proposed Converter

In this subsection, the voltage and current stresses of all the active and passive devices are evaluated. For brevity, the calculations are not shown in the current version of the paper. However, all the design equations are tabulated in Tables 4.3 and 4.4.

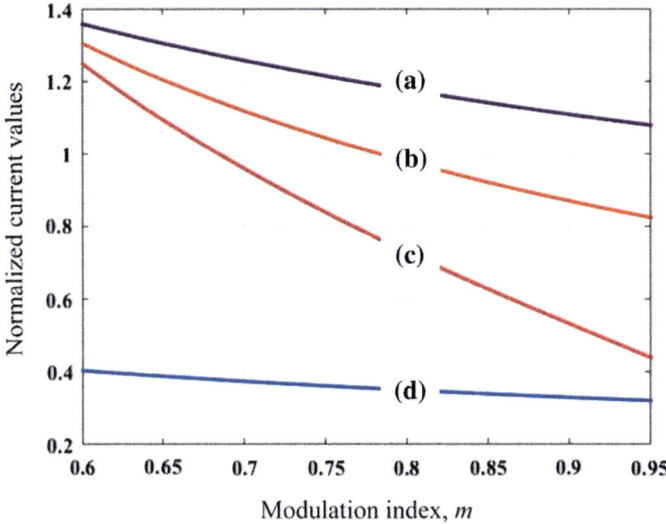

Fig. 4.11 *rms* Current stress calculation using design equations. **a** Boost diode, D_b. **b** CDR diode, D_1 and D_2. **c** Boost MOSFET, M_1. **d** Matrix switches, $S_1 – S_6$

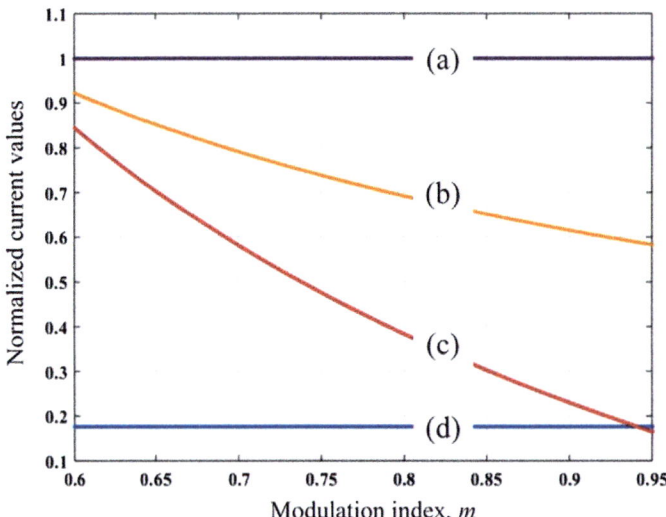

Fig. 4.12 Current stress calculation using design equations. **a** Boost diode, D_b. **b** CDR diode, D_1 and D_2. **c** Boost MOSFET, M_1. **d** Matrix switches, $S_1 – S_6$

Table 4.5 Theoretical and simulation values of current for $G_v = 1.17$ at $m = 0.75$

Currents	RMS		AVG	
	Theo.	Simul.	Theo	Simul.
Matrix Switch	2.4322	2.417	1.1884	1.181
Boost MOSFET	5.649	5.69	3.205	3.246
Boost Diode	8.197	8.153	6.75	6.734
CDR Diodes	7.03	6.836	4.977	4.996

Figures 4.11 and 4.12 show the normalized rms and avg value of currents in the active and passive devices of the converter. With increase in modulation index, m, the rms and avg value of the currents decrease. To validate the design equations, the currents stress of semiconductor devices are theoretically calculated for 1/4th of the specifications shown in Table 4.7 and subsequently, compared with the results obtained through the digital simulation. The values are tabulated in Table 4.5 which show excellent agreement with each other.

4.6   Comparative Evaluation of the Proposed Converter

The boost-buck converter presented in [5, 6] is configured as a Vienna rectifier followed by a buck converter as shown in Fig. 4.13. The output voltage of Vienna rectifier is boosted DC voltage which is stepped down using a buck stage to the desired output voltage. As shown in Table 4.6, the Vienna rectifier uses less number of active switches (5 in total as opposed to 13 in the proposed converter) compared to the proposed topology. However, the number of diodes are significantly higher (20 as opposed to 3 in the proposed topology). Moreover, in the proposed topology, each switch is implemented by connecting two MOSFETs back to back which are driven by a common gate signal. Therefore, the number of gate drivers required are almost same.

The boost-buck rectifier requires large inductor in the input side. Moreover, the boost type of rectifiers have limit over minimum output DC voltage. Therefore, the intermediate DC voltage of the boost-buck rectifier is high which increases the voltage rating of the semiconductor devices. For example, for 230 V rms AC, the minimum intermediate DC voltage is 650 V DC. In [6] the boost buck and buck-boost type of rectifiers have compared in terms of power density, it has been found that the buck-boost type of rectifiers provide higher power density compared to the boost-buck type. One of the main reasons is reduced size of filter (inductor size/volume) in the buck-boost than the boost-buck.

In terms of control, the boost type of rectifier has more complexity in sensing and control implementation. The control of boost-buck rectifier requires three current sensors whereas the proposed converter simply requires one DC sided current sensor

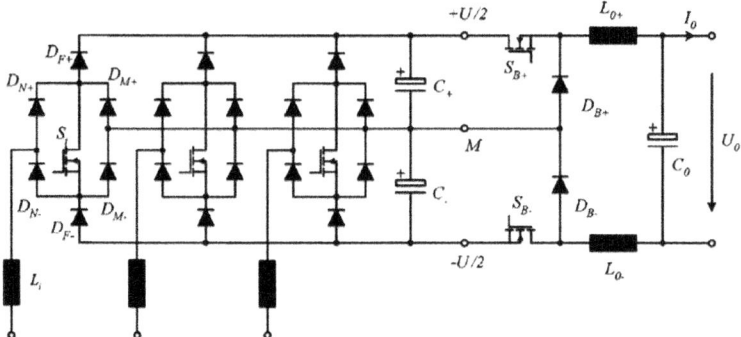

Fig. 4.13 Circuit schematic of the Vienna rectifier based three phase boost-buck rectifier proposed in [5, 6]

Table 4.6 Comparison of the proposed converter with Vienna rectifier based boost buck converter

Converter	Type	No. of switches	No. of diodes	No. gate drivers
Refs. [5, 6], Fig. 4.13	Boost-buck	5	20	5
Proposed converter	Buck-boost	13	3	7

contributing to lower control realization effort. Further, the proposed buck-boost does allow direct start up unlike the boost-buck rectifier which requires pre charging of the DC link capacitor. In the boost-buck topology, an additional control is employed for balancing the voltage across the intermediate capacitors. Even though the number of active switches in proposed buck-boost topology is higher than the boost-buck topology, other aspects such as significantly lesser number of diodes, almost equal gate drive requirements, partial ZVS in switches, lower sensing effort, higher power density and the soft-start capability make the proposed converter a potential choice for applications such as MEA where *power density* and *reliability* are of high concern.

4.7   Experimental Results and Discussion

4.7.1   Prototype Specifications

To demonstrate the suitability of the converter, a scale down hardware prototype is designed and developed and experimental tests are performed. The prototype converter is tested for 25% of the specification shown in Table 4.7. The specifications are chosen for voltage gain, $G_v = 1.17$. The design parameters of the converter are shown in Table 4.8. For experiment at scale down power, the input voltage and the

Table 4.7 Specifications of the proposed converter

Parameter	Value
Three phase input voltage, $V_{in,rms}$	230 V, 400 Hz
Output voltage, V_o	270 VDC
Output power, P_o	2 kW

Table 4.8 Design parameters of the proposed converter

Parameter	Value
Input filter inductors, (L_a, L_b, L_c)	200 μH
Input filter capacitors, (C_a, C_b, C_c)	1.2 μF
output filter inductors, L_{f1}, L_{f2}	2.5 mH
Output filter capacitor, C_o	400 μF
Switching frequency of the matrix converter, f_s	40 kHz
Load resistance, R_o	10 Ω
Switching frequency of the boost converter, f_{sb}	150 kHz

Table 4.9 Selected components for the proposed converter

Components	Part number
MOSFETs for the matrix	C3M0065090D
CDR Diodes (D_1, D_2)	IDP15E65D2
Boost MOSFET, M1	C3M0065090D
Boost diode	VS-20ETF06FPPb
Input filter inductor, L_a, L_b, L_c	CTX22-16885
Input filter capacitor, C_a, C_b, C_c	ECW-FD2W125J
Output filter inductor, L_{f1}, L_{f2}	159ZL
Output filter capacitor, C_o	EKMQ401VSN331MR35S

output voltage are correspondingly scaled down to have better understanding of input power quality and power conversion efficiency of the proposed topology even at this power. The components selected for developing the example converter is shown in Table 4.9. The matrix switches are implemented by back to back connection of SiC MOSFET as shown in Fig. 4.4. The Boost MOSFET is also implemented using SiC. The experimental hardware setup is shown in Fig. 4.14.

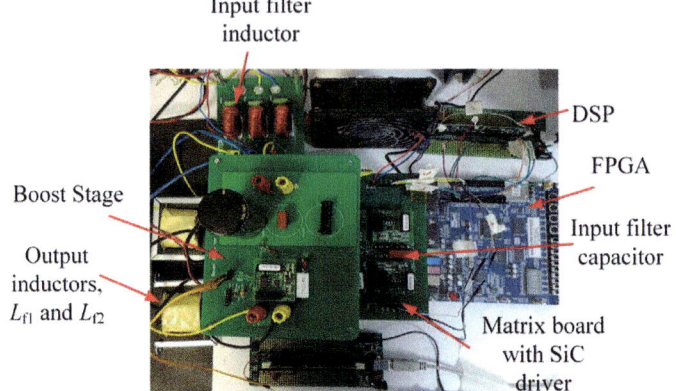

Fig. 4.14 Experimental hardware setup of the proposed converter

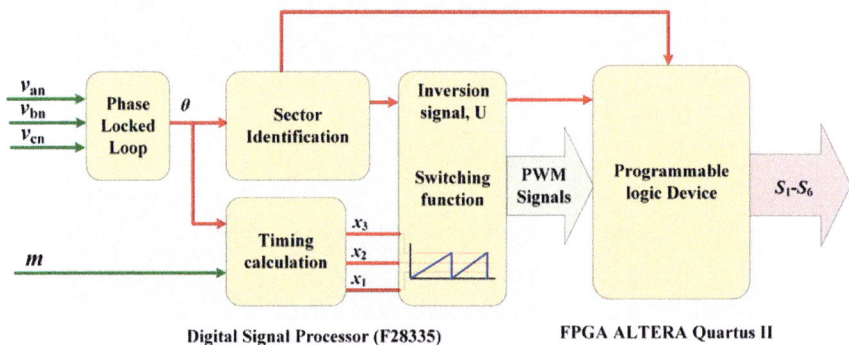

Fig. 4.15 Digital implementation of the proposed switching scheme for the matrix converter

4.7.2 Digital Switching Scheme Implementation

The presented SVM based modulation scheme is implemented using DSP (F28335 TI) and FPGA Quartus II. The block diagram of the digital implementation is shown in Fig. 4.15. The three phase input voltages are sensed and given to Phase Locked Loop (PLL). The PLL generates the angle θ which is used for sector identification and timing calculation. Based on the angle θ and timing values, PWM signals are generated using DSP which is subsequently processed using FPGA with sector information to provide final switching signals. It is important to note here that FPGA only carries out complex boolean operation and thus can be easily replaced with low cost Complex Programmable Logic Device (CPLD) resulting in more efficient and cost-effective control implementation. Figure 4.16 shows the FPGA output signal for the three adjacent matrix switches. A dead time, t_d is provided between adjacent

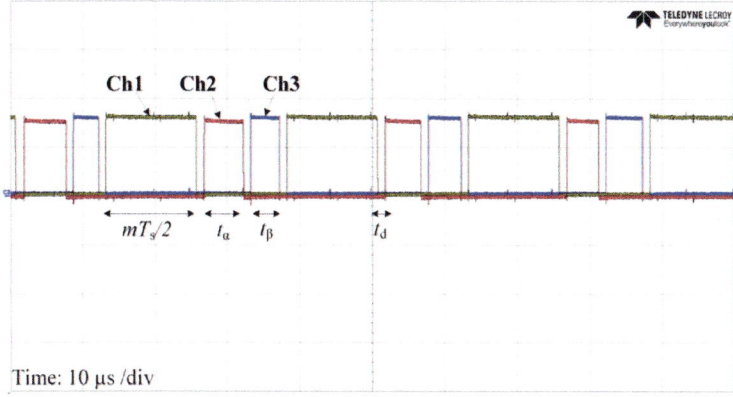

Fig. 4.16 Switching signal for the three adjacent matrix switches, S_1, S_3 and S_5. Ch1: 2 V/div, Ch2: 2 V/div Ch3: 2 V/div

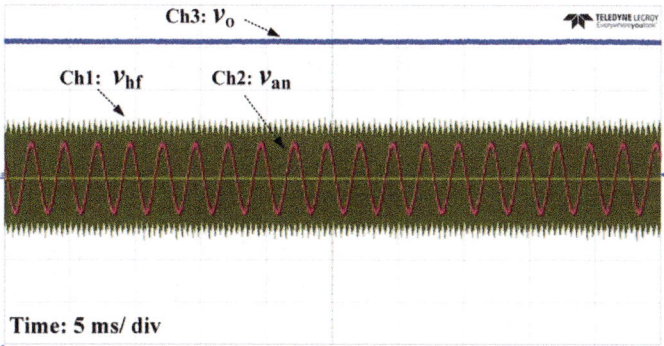

Fig. 4.17 Ch1: High frequency ac voltage (100 V/div), v_{hf}. Ch2: Phase-a voltage, v_{an} (100 V/div). Ch3: Output DC voltage, v_o (20 V/div)

switches for avoiding short circuit of the input filter capacitor. The details of digital implementation of the modulation scheme for the matrix topology is provided in Chap. 3.

4.7.3 Experimental Results

The matrix topology is used to convert line frequency ac voltages into single phase high frequency ac voltage. Figure 4.17 shows the phase-a voltage, v_{an}, high frequency ac voltage, v_{hf} and output DC voltage, v_o. The 400 Hz ac three phase ac voltages are converted into 40 kHz single phase bipolar symmetrical high frequency ac voltage. The proposed converter is tested with output load resistance, $R_o = 10\ \Omega$. The

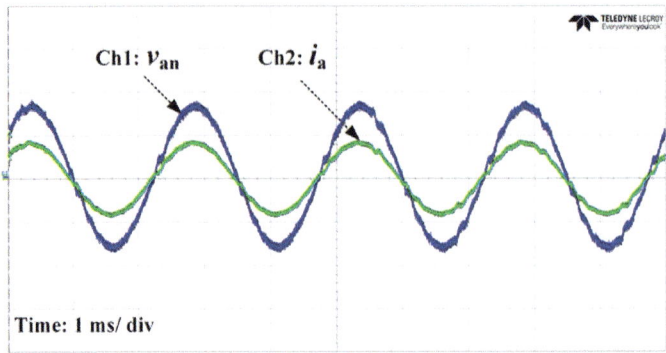

Fig. 4.18   Ch1: Phase- a voltage, v_{an} (50 V/div). Ch2: Phase-a current, i_a (5 A/div)

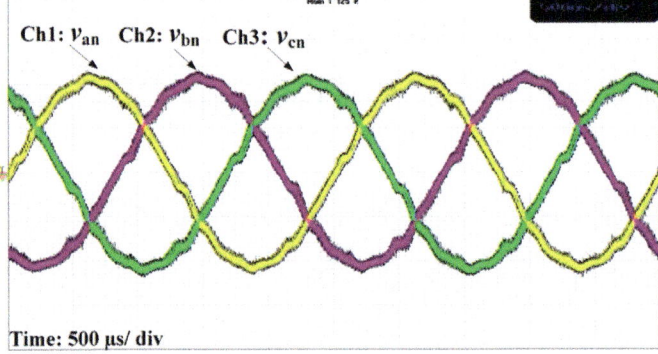

Fig. 4.19   Three phase input ac current, Ch1: i_a (2 A/div), Ch2: i_b (2 A/div) and Ch3: i_c (2 A/div)

phase-a voltage, v_{an} and phase-a current, i_a are shown in Fig. 4.18. The displacement power factor is found to be almost unity. Figure 4.19 shows three phase ac current. The input currents in all the three phases are balanced and symmetrical. Figure 4.20 shows the high frequency ac voltage, v_{hf}. The high frequency ac voltage is synthesized using SVM and alternatively inverted using inversion signal, U to generate symmetrical bipolar ac voltage. In Fig. 4.21, the high frequency ac voltage, v_{hf} and voltage across boost MOSFET, v_{sw} are shown. A dead time is provided during the switching transition from one leg to other leg in the matrix converter to avoid the short circuiting of the input filter capacitor. Figure 4.22 shows voltage across one of the MOSFETs of the matrix switch. It is worth noticing that the voltage across the MOSFETs is zero for $\frac{\pi}{3}$ duration resulting in ZVS of the MOSFET. Further, the THD of the input current at full load is experimentally observed. The THD of the input current is found to be 3.2% at full load.

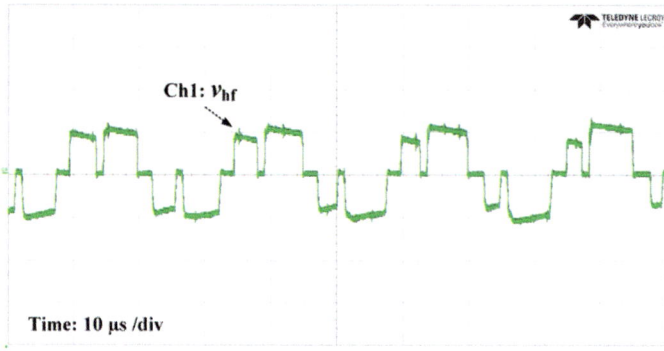

Fig. 4.20 Ch1: High frequency ac voltage, v_{hf} (100 V/div)

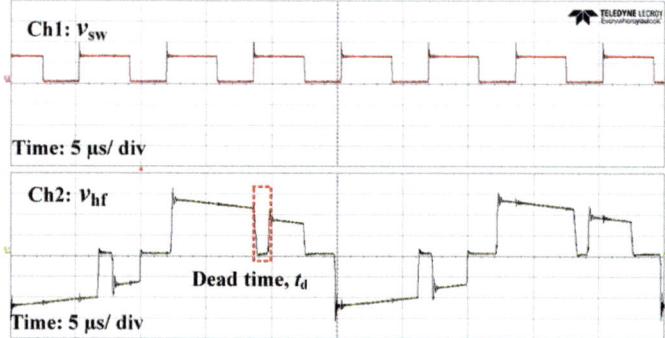

Fig. 4.21 Ch1: Voltage across the boost MOSFET, v_{sw} (50 V/div). Ch2: High frequency ac voltage, v_{hf} (50 V/div)

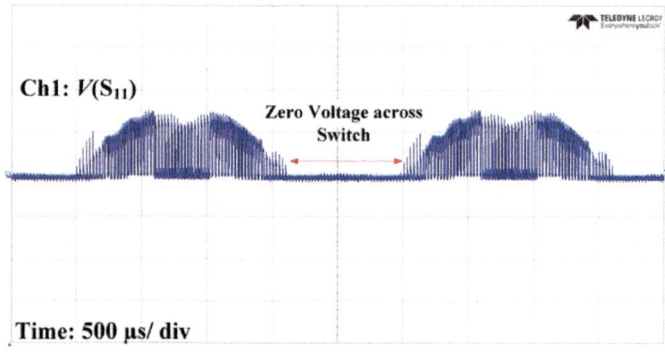

Fig. 4.22 Ch1: Voltage across one of the MOSFETs of the matrix switch (100 V/div), $V(S_{11})$

4.7.4   Discussion on the Power Conversion Efficiency of the Proposed Converter

For the hardware prototype of the proposed converter, SiC, 900 V MOSFETs are used for matrix and boost switch implementation. As the current in matrix switches passes through the two back to back connected MOSFETs, it is imperative to have MOSFETs with low on resistance to reduce the switch conduction loss. Further, the CDR diode carries significant amount of high frequency current. The CDR diodes are chosen for low forward voltage drop and low reverse recovery charge. The two output filter inductors are required to have low resistance to reduce losses. The theoretical loss is calculated for the proposed converter at full load ($R_o = 10\,\Omega$). The loss distribution is shown in Fig. 4.23. The theoretical power conversion efficiency of the converter is found to be 88.96% which is very close to the experimental efficiency of 86.12%. It is worth mentioning that the input/output voltage/current ripples are ignored for simplicity while calculating for the total loss of the proposed converter. A R-C snubber (20 Ω and 1 nF) circuit is used to reduced the high frequency ringing across the boost MOSFET. Additionally, a R-C snubber (50 Ω and 0.5 nF) is put across each matrix switch to reducing high frequency voltage ringing. As the switch current always flows through the SiC MOSFETs channel which exhibits very low ON resistance, the proposed topology promises very low conduction loss. Moreover, each of the MOSFETs matrix partially get ZVS (1/3 of the total mains period), the switching loss are also minimized. With increase in the modulation index, both *rms* and *avg* currents in semiconductor reduce and therefore, the efficiency of the

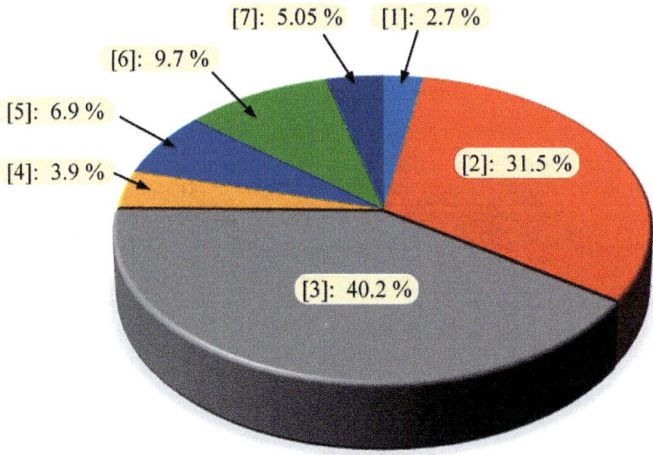

Total loss : 56.45 W

Fig. 4.23 Total theoretical loss distribution of the converter. Total loss is found to be 56.45 W at full load. [1]. Input filter loss [2]. Matrix switch loss [3]. CDR diodes loss [4]. CDR Inductor loss [5]. Boost MOSFET loss [6]. Boost diode loss [7]. Snubber loss

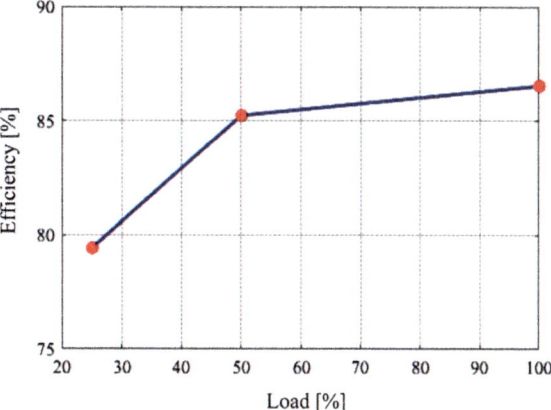

Fig. 4.24 Experimental efficiency of the scaled down converter prototype at different load

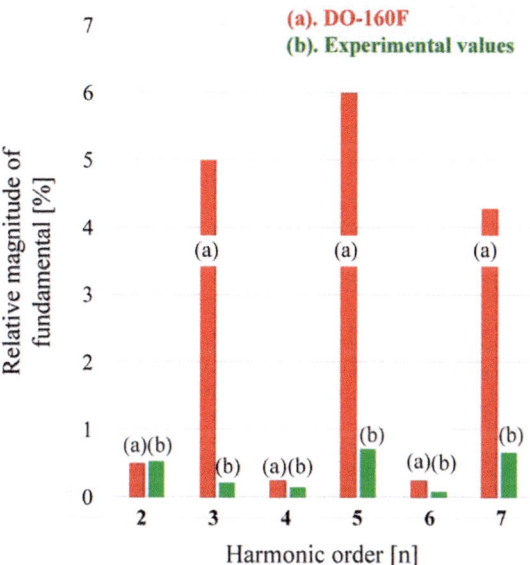

Fig. 4.25 Current harmonic distribution of the input current. The experimental current harmonics shown in blue are compared with DO-160F standard shown in red color

proposed topology can be further improved by operating it at higher modulation index. The efficiency of the scaled down converter is calculated experimentally for 100–25% load variation. Figure 4.24 shows the efficiency of the converter for different loads. At lower load, the efficiency of the converter drops due to constant losses such as snubber losses. Additionally, the input filter capacitors takes finite amount of reactive power which further reduces the overall power conversion efficiency at lower loads.

Fig. 4.26 **a** Displacement Power Factor (DPF) versus output load. **b** Total Harmonic Distortion (THD$_i$) of the input current versus output load

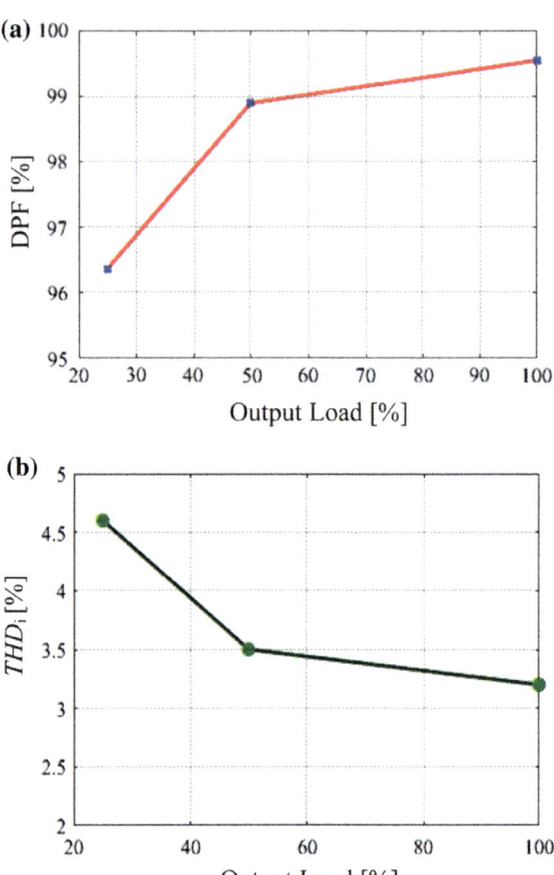

4.7.5   Discussions on the Input Power Quality of the Proposed Converter

In this subsection the input power quality of the proposed converter is discusses. The THD of the input current for the scaled down prototype at full load is calculated and its harmonic distribution is compared with standard limit specified by DO-160 which is shown in Fig. 4.25. The individual current harmonics are within limits. The Displacement Power Factor (DPF) of the converter is experimentally checked at different output load and shown in Fig. 4.26a. At full load the converter operates with unity power factor. Even at 50% of the full load, the DPF is close to 99%. The input current THD is also plotted for different output load as shown in Fig. 4.26b. The overall THD is found to be 3.2% at full load. The proposed converter provides superior THD. Even at 25% of the load, the THD is less than 5%. It should be noted that by selecting smaller value of the filter capacitors, the DPF variation for decreasing load can be further narrowed. However, the THD of the input increases for low values of input filter capacitor. Therefore, a trade off is needed between DPF for low load and input current THD.

4.8 Conclusion

A new matrix based non-isolated three phase buck-boost converter suitable for aircraft systems is presented in this paper. The proposed converter is able to provide voltage gain for which the conventional three phase PWM ac–dc converter such as buck, boost and buck-boost are not suitable. The steady state analysis and design of the proposed converter are provided in details. The SVM based modulation scheme is digitally implemented for 40 kHz high frequency AC output from the matrix converter. The matrix switches are implemented through state-of-art SiC switches and experimental efficiency of 86.12% is obtained at 455 W output power. Even at 25% of the full load, the power conversion efficiency of the proposed converter is found to be 79%. The scale down prototype demonstrates superior input power quality with THD = 3.2%. Moreover, the DPF is unity at full load. For 25% of the load, the proposed converter shows the input current THD better than 5% and DPF more than 96%. Additionally, the input current harmonics are compared with DO-160F harmonic limit and it is found to be within limits. The benefits of and limitations of the converter can be summarized as follows:

Benefits:

1. The proposed converter topology allows converter to be operated at higher modulation index unlike the presented buck-boost converter in [5, 6].

2. Each of the matrix switches are controlled by single control which reduces the control complexity. The inductor currents freewheels through CDR diodes when all the matrix switches are open and thus, current commutation problem is circumvented.

3. The presented SVM based switching scheme does not require body diode conduction and therefore, the losses associated with body diode- *conduction loss and reverse recovery loss* are eliminated. Moreover, SiC MOSFETs have lower on resistance which reduces switch conduction loss. Moreover, each of the MOSFETs in the matrix topology achieves partial ZVS resulting in reduced switching loss.

Limitations:

1. The proposed topology has significantly higher number of MOSFETs.

2. The high switching frequency operation requires good PCB layout design. The bidirectional matrix switch should be implemented for low parasitic.

New research in the development of monolithic bidirectional switch [8] promises improved switching characteristic and further performance improvement of the proposed topology in the near future.

References

1. J. Brombach, M. Jordan, F. Grumm, D. Schulz, Converter topology analysis for aircraft application, in *International Symposium on Power Electronics Power Electronics, Electrical Drives, Automation and Motion* (2012), pp. 446–451
2. J. Brombach, A. Lcken, B. Nya, M. Johannsen, D. Schulz, Comparison of different electrical hvdc-architectures for aircraft application, in *Electrical Systems for Aircraft, Railway and Ship Propulsion (ESARS), 2012* (2012), pp. 1–6
3. U. Borovi, Analysis and comparison of different active rectifier topologies for avionic specifications. Master's thesis, Universidad Politcnica de Madrid, the Spain (2014)
4. J. Kolar, T. Friedli, The essence of three-phase pfc rectifier systems-part i. IEEE Trans. Power Electron. **28**, 176–198 (2013)
5. T. Nussbaumer, J.W. Kolar, Comparison of 3-phase wide output voltage range pwm rectifiers. IEEE Trans. Ind. Electron. **54**, 3422–3425 (2007)
6. Y. Nishida, J. Miniboeck, S.D. Round, J.W. Kolar, A new 3-phase buck-boost unity power factor rectifier with two independently controlled dc outputs, in *APEC 07 - Twenty-Second Annual IEEE Applied Power Electronics Conference and Exposition* (2007), pp. 172–178
7. J. Kikuchi, T.A. Lipo, Three-phase pwm boost-buck rectifiers with power-regenerating capability. IEEE Trans. Ind. Appl. **38**, 1361–1369 (2002)
8. R. Sittig, A. Krysiak, S. Chmielus, Monolithic bidirectional switches promise superior characteristics, in *2004 IEEE 35th Annual Power Electronics Specialists Conference (IEEE Cat. No.04CH37551)*, vol. 4 (2004), pp. 2977–2982

Chapter 5
A SQR Based High Voltage LLC Resonant DC–DC Converter

5.1 Introduction

This chapter mainly focuses on analysis and design of HV resonant DC–DC converter for powering Traveling Wave Tube (TWT) in MPM based transmitters. Due to low weight/volume, MPM based transmitters are especially suited for smaller aircrafts such as UAVs for different applications including surveillance and navigational purposes. A brief review of HV DC–DC converter is given in Sect. 5.2. The new contribution of the chapter is highlighted in Sect. 5.3. In Sect. 5.4, the topology and operation of the proposed converter are discussed in details. The comprehensive steady state analysis and design of the converter are carried out in Sects. 5.5 and 5.6, respectively. The design of the converter is verified by simulation and experimental results in Sect. 5.7. Moreover, the Sect. 5.7 also presents the brief discussion on the power conversion efficiency and comparative evaluation of the two design methods-the proposed method and First Harmonic Approximation (FHA). Section 5.8 provides the conclusion.

5.2 A Brief Review of HV DC–DC Converter

Several high voltage DC–DC converters are presented in literature. High switching frequency is essential for reducing the size of magnetic and passive elements. Resonant converters are used for operation where very high switching frequency are required because of their natural ability to switch softly between the ON and OFF states [1–5]. For a high voltage DC–DC converter, the effects of parasitic elements become quite significant, especially at higher switching frequencies. However, in resonant converters these parasitics can be suitably utilized by forming a resonant tank. The requirement of large turns-ratio for the transformer in a high voltage DC–DC converter increases the impact of parasitics and therefore, restricts the maximum

© Springer Nature Singapore Pte Ltd. 2018
A. K. Singh, *Analysis and Design of Power Converter Topologies for Application in Future More Electric Aircraft*, Springer Theses, https://doi.org/10.1007/978-981-10-8213-9_5

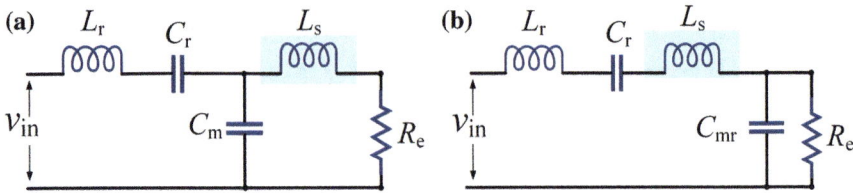

Fig. 5.1 **a** Equivalent LCC circuit when parallel capacitor is used in primary side. **b** Equivalent LCC circuit when parallel capacitor is used in secondary side

achievable frequency [2, 6]. Another disadvantage of the large turns-ratio in a high frequency transformer is increased size and volume of the transformer as more number of turns requires large window area. The high voltage transformer needs sufficient isolation between the primary and secondary windings which, essentially, increases the insulation quantity contributing to increased cost. Moreover, the increased insulation in the windings further exacerbates the parasitic effects leading to poorer performance of the transformer at high switching frequency. In the literature [7–12] various kind of voltage multipliers have been used to design the high voltage DC–DC converter. The combination of the series and parallel resonant tank, LCC tank is much popular and widely used to generate high voltage [4, 13–16]. The LCC tank can provide voltage gain greater than unity with a lesser variation in the switching frequency for regulating the output voltage for different loads. However, the performance of the LCC tank deteriorates over wide input voltage range because of the large circulating energy and turn off current of the MOSFETs. Another key drawback of the LCC tank is that it requires a physical capacitor for the parallel resonant component. When the parallel capacitor is placed on the primary side, then the LCC tank degenerates into a LCC-L tank where the second inductor arises from the transformer leakage inductance as shown Fig. 5.1a [17]. Typically in a high voltage high turns ratio transformer, the leakage inductance can be quite high which deteriorates the voltage gain of the LCC tank. Therefore, the parallel capacitor of a LCC tank should be placed in the secondary side in order to avoid the high leakage inductance arising in the high voltage transformers as shown in Fig. 5.1b; however, in a high output voltage DC–DC converter, this may not be a viable option because an extremely high voltage high frequency capacitor will be required to implement the secondary side parallel resonant capacitor.

Figure 5.2 shows the derivation of SQR circuit using traditional 2-stage Cockcroft voltage multiplier. By splitting the Cockcroft voltage multiplier by half and combining it in parallel, the SQR circuit can be realized. Due to parallel combination, the output impedance of SQR circuit is lower than the Cockcroft voltage multiplier. Moreover, the frequency of the output voltage ripple in SQR circuit is two times of the Cockcroft voltage multiplier. The benefits of using SQR over other voltage multiplier circuits are described in [18].

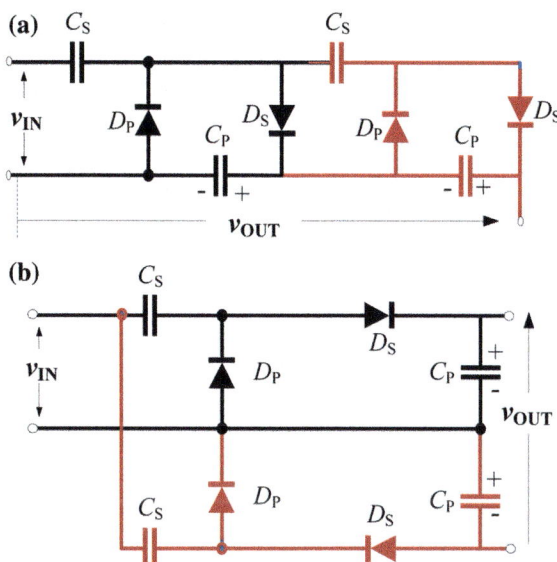

Fig. 5.2 a Two stage Cockcroft voltage multiplier. **b** Derived Symmetrical Quadrupler Circuit using Cockcroft voltage multiplier

5.3   The New Contributions of the Chapter

An LLC converter provides significant advantage in terms of power conversion efficiency and power density for a wide input voltage range [5, 19–21]. Moreover, operation of the LLC tank at switching frequency lower than the series resonant frequency results in smaller tank elements for a given switching frequency. Operating the SQR based high voltage LLC resonant DC–DC converter at switching frequency lower than the series resonant frequency of the LLC tank makes the secondary current discontinuous and therefore, the usual analysis based on the FHA is not accurate [22]. This chapter provides a comprehensive analysis of the LLC tank along with the SQR by solving the basic differential equation in the different modes of operation which enables more accurate and realistic design for the converter. In this chapter, the step by step design procedure of the converter along with various design trade-off is discussed in details. Moreover, a comparison in the accuracy of output voltage ripple estimation is provided to demonstrate the efficacy of the proposed analysis and design method. In brief, the main contributions of this chapter are as follows:

- Comprehensive analysis and design of the combination of two topologies, an *LLC* resonant full bridge inverter and a Symmetrical Quadrupler Rectifier (SQR) for an additional voltage gain from the resonant tank;
- proposed a new differential equation based analysis method and demonstrated its accuracy in comparison to the usual FHA method which is subsequently validated through simulation and experimental results;
- proper analysis of the SQR using FHA which has never been reported in literature for an *LLC* resonant converter; and

- the unique characteristics of the SQR diode currents such as discontinuity in the current, the effect of the ratio of the parallel and series capacitor on the voltage and current stresses of the SQR diodes are explained in details.

5.4 Topology and Operation of the Proposed HV DC–DC Converter

In this section, the topology of the converter is explained. subsequently, the different modes of operation of the converter is explained. Figure 5.3 illustrates the circuit schematic of the converter. The front-end of the converter consists of a voltage fed full-bridge inverter followed by an *LLC* resonant tank and a high frequency transformer. At the output side, the SQR circuit is used. Devices S_1, S_2, S_3 and S_4 are used to implement the full-bridge inverter followed by the series resonant inductor L_r and the series resonant capacitor C_r giving rise to the primary side series resonant tank connected to the high frequency transformer T with turns ratio $1 : N$. At the secondary side of the transformer, the SQR circuit is used for generating the high voltage DC, V_o with a voltage gain of four. The SQR is formed by capacitors, C_5–C_8 and by diodes, D_5–D_8. Since the capacitors C_7 and C_8 are parallel to the load, they are termed as parallel capacitors of the SQR. Similarly, the capacitors, C_5 and C_6 are termed as series capacitors.

Figure 5.4 shows the theoretical waveforms of the presented high voltage converter during one switching cycle. The complete operation of the LLC resonant tank with the SQR can be divided into twelve modes. Because of symmetrical operation during

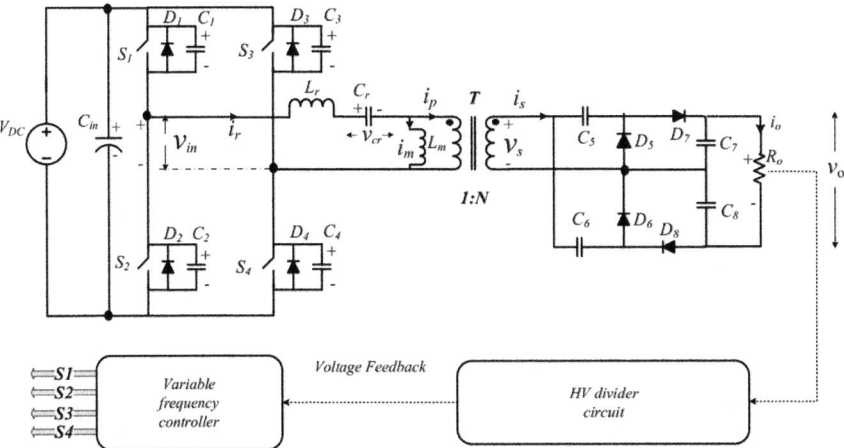

Fig. 5.3 Circuit schematic of the SQR based high voltage DC–DC converter. The total high voltage gain is shared among LLC tank (operating below series resonant frequency), high frequency transformer and SQR (4 × gain)

Fig. 5.4 Theoretical waveforms of the converter. **a** shows inverter voltage v_p, resonant current i_r, and magnetizing current i_m. **b** shows the secondary current which is discontinuous for time t_d verifying the operation of converter below series resonant frequency f_r. **c** show secondary transformer voltage v_s. **d** shows the diode currents in SQR circuit. **e** is the converter output voltage v_0

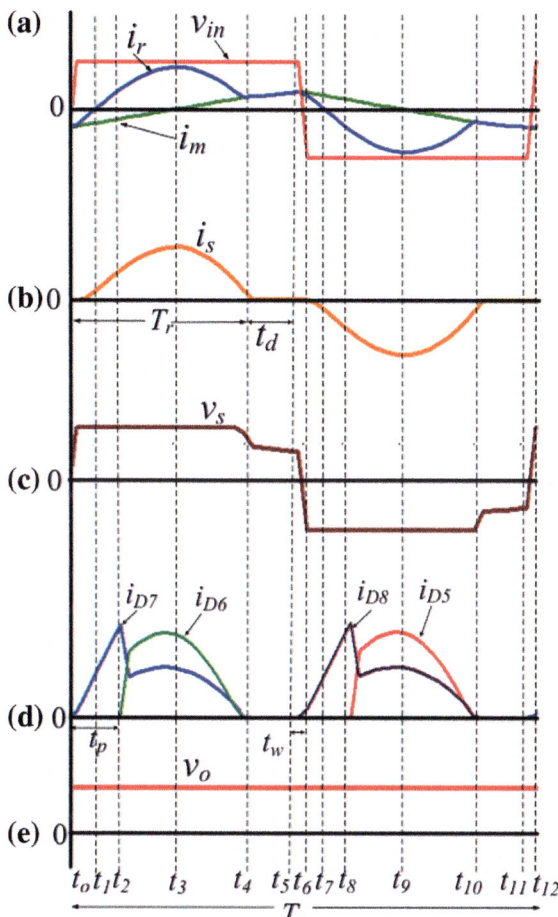

the positive and negative half cycle of the operation, only six modes of the operation are explained. The operation of the converter for *mode*-1 to *mode*-6 has been shown in Figs. 5.5 and 5.6. To get the additional voltage gain from the LLC tank, the converter is operated at switching frequency lower than the series resonant frequency, f_r. Followings are the assumptions made for explaining the different modes of operation for the presented SQR based high voltage DC–DC converter.

- The ON resistance of MOSFETs (S_{1-4}), R_{ds} and body diodes (D_{1-4}) forward voltage drop, V_{sdf} are assumed to be zero;
- the forward voltage drop, V_f of the SQR diodes is zero;
- the output load current, i_o is constant and is equal to I_o;
- the DC output voltage, v_o is assumed to be ripple free and is equal to V_o;
- the secondary current i_s is divided into two parts- i_{s1} and i_{s2}. i_{s1} flows through diode D_5 (during −ve half cycle) or diode D_6 (during +ve half cycle) whereas

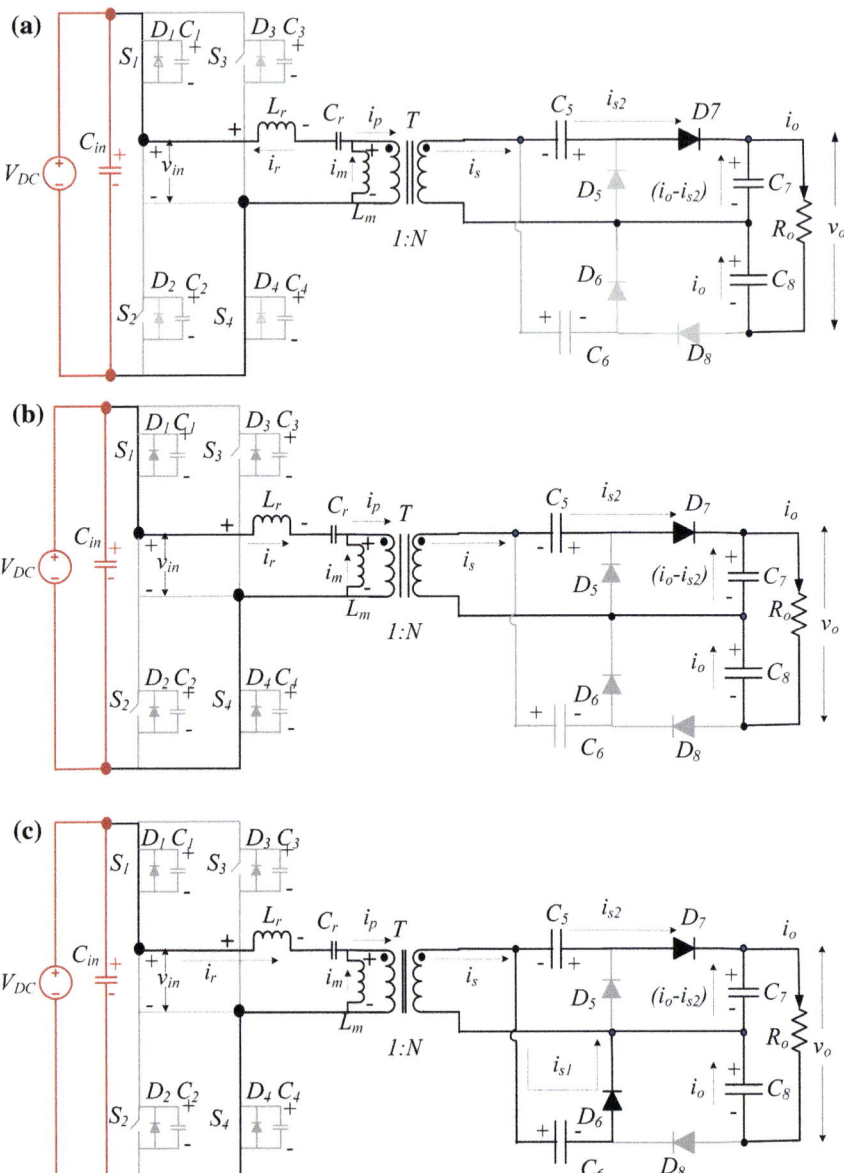

Fig. 5.5 Modes of operation of the SQR based high voltage DC–DC converter. **a** Mode-1 **b** Mode-2 **c** Mode-3

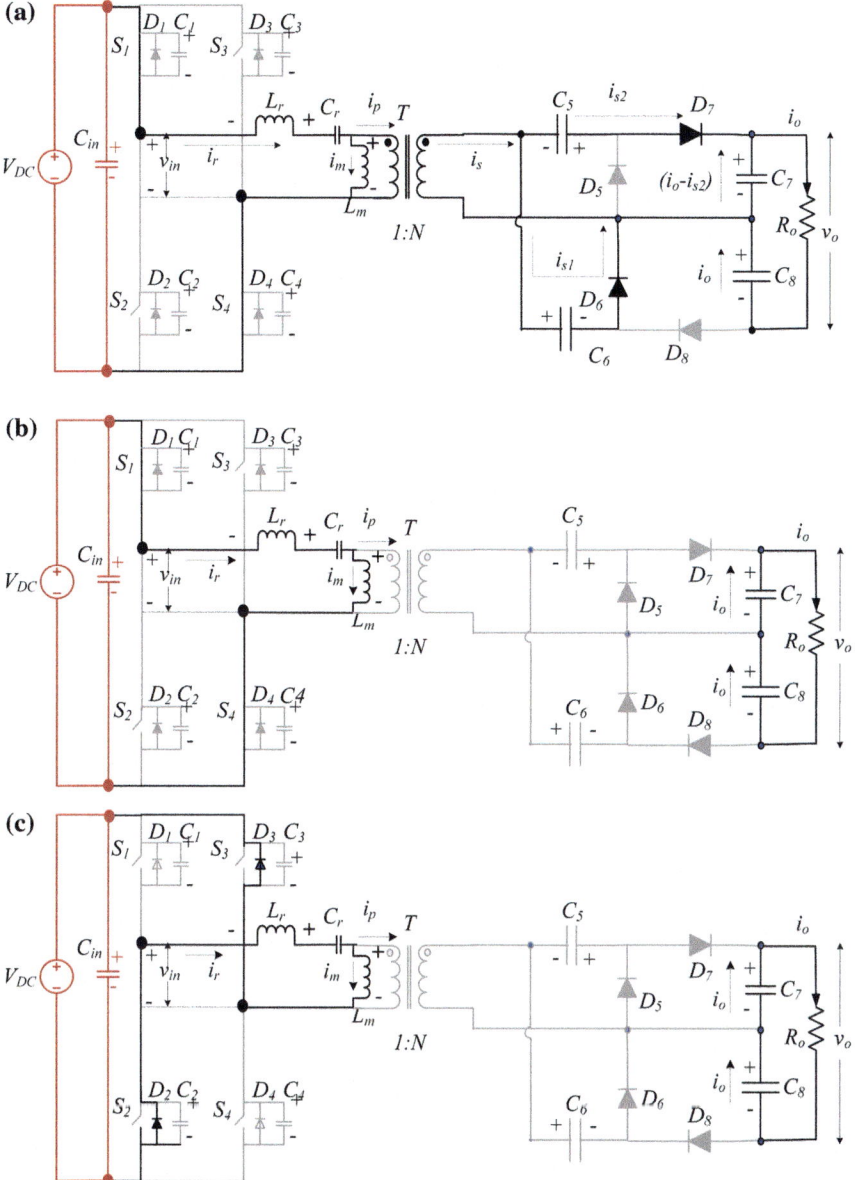

Fig. 5.6 Modes of operation of the SQR based high voltage DC–DC converter. **a** Mode-4 **b** Mode-5 **c** Mode-6

i_{s2} flows through diode D_7 (during +ve half cycle) or diode D_8 (during −ve half cycle);

- $C_8 = C_7; C_5 = C_6$

The symmetrical operation of the SQR circuit in both positive and negative cycle required the parallel capacitors, C_7 and C_8 to be equal. Similarly, the series capacitors, C_5 and C_6 should be equal. If $C_8 = C_7 = C_p$ and $C_5 = C_6 = C_s$ then, the ratio of the values of the parallel capacitor and series capacitor is defined as α.

$\alpha = \frac{C_p}{C_s} = \frac{C_7}{C_5} = \frac{C_8}{C_6}$.

5.4.1   Mode-1

$t_0 \leq t \leq t_1$: During this mode of the operation, switches S_1 and S_4 are ON giving rise to $v_{in} = V_{DC}$ as shown in Fig. 5.4a. At $t = t_0$, the series resonant current, i_r and the magnetizing current, i_m are equal. However, the rate of rise of i_m is lower than the rate of rise of i_r which results in net positive transformer primary current, i_p. During this mode, the tank resonates with the series resonant frequency, f_r given as,

$$f_r = \frac{\omega_r}{2\pi} = \frac{1}{2\pi\sqrt{L_r C_r}}. \tag{5.1}$$

The characteristic impedance of the tank can also be defined as

$$Z_c = \sqrt{\frac{L_r}{C_r}}. \tag{5.2}$$

If the voltage across the magnetizing inductor is v_p then with the above assumptions,

$$v_p(t - t_o) = \frac{V_0}{4N}. \tag{5.3}$$

where, N is the turns ratio of the transformer. The voltage across the magnetizing inductance, L_m can be described as,

$$v_p(t - t_o) = L_m \frac{di_m}{dt}. \tag{5.4}$$

The voltage across the series elements of the LLC tank, L_r and C_r is $(v_{in} - v_p)$, which can be described as the following equation,

$$v_{in}(t - t_o) - v_p(t - t_o) = L_r \frac{di_r}{dt} + \frac{1}{C_r} \int_{t_o}^{t} i_r(t) dt. \tag{5.5}$$

where, $v_{in} = V_{DC}$.

The transformer secondary current, i_s is equal to i_{s2} which flows through diode D_7 discharging and charging the capacitors, C_5 and C_7 respectively. The transformer secondary voltage, v_s during this mode is given by

$$v_s(t - t_o) = v_{C_7}(t - t_o) - v_{C_5}(t - t_o) \tag{5.6}$$

During this mode, v_s remains lower than the voltage across capacitor C_6, $V(C_6)$ which makes diode D_6 reverse biased. As D_6 does not conduct, i_{s1} remains zero during this mode. if the change in voltage across SQR capacitors is defined by Δ, then

$$\Delta v_{C_5}(t - t_o) = -\frac{1}{C_5} \int_{t_o}^{t} i_{s2}(t)dt. \tag{5.7}$$

$$\Delta v_{C_6}(t - t_o) = 0. \tag{5.8}$$

$$\Delta v_{C_7}(t - t_o) = \frac{1}{C_7} \int_{t_o}^{t} (i_{s2}(t) - i_o(t))dt. \tag{5.9}$$

$$\Delta v_{C_8}(t - t_o) = -\frac{1}{C_8} \int_{t_o}^{t} i_o(t)dt. \tag{5.10}$$

This mode ends when the series resonant current, i_r becomes zero.

5.4.2 Mode-2

$t_1 \leq t \leq t_2$: This mode is exactly similar to mode-1 with an exception that i_r becomes positive as shown in Fig. 5.4b. The equation described in mode-1 will be same for mode-2. This mode ends when secondary voltage, v_s becomes equal to voltage across capacitor C_6, V_{C_6} which results in the turn ON of the diode, D_6.

5.4.3 Mode-3

$t_2 \leq t \leq t_3$: During this mode of operation, the operation in the primary side of the transformer remain similar to mode-2. However, in the secondary side, v_s becomes equal to V_{C_6} which results in the conduction of diode D_6. The current, i_{s1} flows through the diode D_6 charging the capacitor C_6. The current i_s is divided into two parts- i_{s1} and i_{s2} which can be described as follows,

$$i_s(t - t_2) = i_{s1}(t - t_2) + i_{s2}(t - t_2) \tag{5.11}$$

$$i_{s1}(t - t_2) = \frac{i_s(t - t_2)(\alpha + 1) - i_o(t - t_2)}{2\alpha + 1} \tag{5.12}$$

$$i_{s2}(t - t_2) = \frac{i_s(t - t_2)\alpha + i_o(t - t_2)}{2\alpha + 1} \tag{5.13}$$

The voltage change in the SQR capacitors can be defined as,

$$\Delta v_{C_5}(t - t_2) = -\frac{1}{C_5} \int_{t_2}^{t} i_{s2}(t)dt \tag{5.14}$$

$$\Delta v_{C_6}(t - t_2) = \frac{1}{C_6} \int_{t_2}^{t} i_{s1}(t)dt \tag{5.15}$$

$$\Delta v_{C_7}(t - t_2) = \frac{1}{C_7} \int_{t_2}^{t} (i_{s2}(t) - i_o(t))dt \tag{5.16}$$

$$\Delta v_{C_8}(t - t_2) = -\frac{1}{C_8} \int_{t_2}^{t} i_o(t)dt \tag{5.17}$$

This mode ends when the magnetizing current, i_m becomes zero.

5.4.4   Mode-4

$t_3 \leq t \leq t_4$: This mode is similar to mode-3 with an exception that the current i_m is positive as shown in Fig. 5.4d. The operation of converter for both primary and secondary side of the transformer is exactly similar. This mode ends when series resonant current, i_r and the magnetizing current, i_m becomes equal again.

5.4.5   Mode-5

$t_4 \leq t \leq t_5$: At the start of this mode, series resonant current, i_r and the magnetizing current, i_m are equal resulting in zero current in the transformer winding ($i_p = 0$). This mode is characterized by time duration, t_d as shown in Fig. 5.4. During this mode, no power is transferred from source to load. The output power is taken from the output capacitors, C_7 and C_8. The operation of the converter is shown in Fig. 5.4e. During this mode, the LLC tank resonates with resonant frequency, f_0 which is given as,

$$f_0 = \frac{\omega_0}{2\pi} = \frac{1}{2\pi\sqrt{(L_r + L_m)C_r}} \tag{5.18}$$

The characteristic impedance of LLC tank during this mode is given as,

$$Z_0 = \sqrt{\frac{(L_r + L_m)}{C_r}} \tag{5.19}$$

The governing circuit equations during this mode is given as,

$$v_p(t - t_4) = (L_r + L_m)\frac{di_r}{dt} + \frac{1}{C_r}\int_{t_4}^{t} i_r(t)dt \tag{5.20}$$

$$i_r(t - t_4) = i_m(t - t_4) \tag{5.21}$$

All of the SQR diodes are turned off during this. The capacitors C_7 and C_8 discharges through load.

$$\Delta v_{C_5}(t - t_4) = 0 \tag{5.22}$$

$$\Delta v_{C_6}(t - t_4) = 0 \tag{5.23}$$

$$\Delta v_{C_7}(t - t_4) = -\frac{1}{C_7}\int_{t_4}^{t} i_o(t)dt \tag{5.24}$$

$$\Delta v_{C_8}(t - t_4) = -\frac{1}{C_8}\int_{t_4}^{t} i_o(t)dt \tag{5.25}$$

This mode ends when the switch S_1 and S_4 are turned OFF.

5.4.6 Mode-6

$t_5 \leq t \leq t_6$: A dead time of t_w duration is provided between switches of same legs ($S_1 - S_2$ and $S_3 - S_4$). All the switches will be off during this mode. However, magnetizing current i_m is still flowing in the resonant tank which flows through the Diode D_2 and D_3 and therefore, discharges the capacitor C_2 and C_3. Assuming that the magnetizing current is large enough to discharge the capacitor, the switches S_1 and S_4 turns ON with zero voltage in the next mode and therefore, achieves ZVS.

ZVS for the switches S_2 and S_3 requires

$$\frac{1}{2}(L_r + L_m)i_m^2(t_5) \geq \frac{8}{3}C_{oss}V_{dc}^2 \tag{5.26}$$

where C_{oss} is drain to source capacitance of the MOSFET. For current i_m, the time duration t_w should be sufficient enough to discharge the capacitors.

$$i_m(t_5)t_w \geq \frac{4}{3}C_{oss}V_{dc} \tag{5.27}$$

The secondary side operation of converter is similar to mode-5. No power is transferred from source to load. The output load is fed by the output capacitors C_7 and C_8. This mode ends when the switches, S_2 and S_3 are turned ON.

With the end of this mode, the first six modes of the operation of the converter during the positive half cycle is finished. For the next six modes, the converter's operation is exactly similar to the first six modes of the operation and can be explained in similar manner.

5.5   Steady State Analysis of the SQR Based High Voltage DC–DC Converter

In this section, the operation of the converter is analyzed depending on the mode of operation described in the Sect. 5.4. The analysis of the converter is divided into seven sections which are explained as follows.

5.5.1   LLC *Resonant Converter in Boost Mode*

The characteristic solutions of the equations derived in Sect. 5.4 for the LLC tank can be solved based on symmetry of the series resonant current, i_r, magnetizing current, i_m and the series capacitor voltage, v_{cr} during half of the switching cycle, $T_s/2$. By symmetry, it can be shown that after half of the switching cycle i_r, i_m and v_{cr} become equal and opposite. Also, the average output power can be computed by averaging the power fed to the load during half of the switching cycle. Consequently, four simultaneous equations can be formed which can be solved to find the solutions of those equations. It is to be noted that during mode-1 to mode-4, the reflected output voltage in the primary side, v_p is assumed to be constant (ignoring the output voltage ripple), therefore characteristic equations for mode-1 to mode-4 remains similar. The solution of the equations from mode-1 to mode-4 is given by

$$i_m(t) = i_m(0) + \frac{V_0}{4NL_m}t \tag{5.28}$$

$$i_r(t) = i_r(0)\cos(\omega_r t) + \frac{V_{DC} - \frac{V_0}{4N} - v_{cr}(0)}{Z_c}\sin(\omega_r t) \tag{5.29}$$

$$v_{cr}(t) = -\left(V_{DC} - \frac{V_0}{4N} - v_{cr}(0)\right)\cos(\omega_r t) +$$
$$Z_c i_r(0)\sin(\omega_r t) + \left(V_{DC} - \frac{V_0}{4N}\right) \quad (5.30)$$

The solution of the equations for mode-5 and mode-6 is given by

$$i_r - i_m = 0 \quad (5.31)$$

$$i_r(t) = i_m(t) = i_r(T_r)\cos(\omega_0(t - T_r)) + \frac{V_{DC} - v_{cr}(T_r)}{Z_0}\sin(\omega_0(t - T_r)) \quad (5.32)$$

The voltage across capacitor is given by,

$$v_{cr}(t) = V_{DC}(1 - \cos(\omega_0(t - T_r))) + Z_0 i_r(T_r)\sin(\omega_0(t - T_r)) + v_{cr}(T_r)\cos(\omega_0(t - T_r)) \quad (5.33)$$

From the symmetry of the voltage and current during half of the switching cycle,

$$v_{cr}(0) = -v_{cr}\left(\frac{T_s}{2}\right) \quad (5.34)$$

$$i_r(0) = -i_r\left(\frac{T_s}{2}\right) \quad (5.35)$$

$$i_m(0) = -i_m\left(\frac{T_s}{2}\right) \quad (5.36)$$

if P_0 is average output power, then it can be given as

$$P_0 = \frac{V_0}{4N}\frac{2}{T_s}\int_0^{T_r}(i_r(t) - i_m(t))dt \quad (5.37)$$

The output power can be further simplified using (5.37) to find the closed form expression as,

$$P_o = \frac{V_o}{2NT_s}\left[\frac{2\sin\beta^2}{Z_c\omega_r}(V_{DC} - v_{cr}(0)) + \frac{V_o}{8L_mN}\left(\beta\omega_r T_s - \frac{\sin(2\beta\omega_r^2 T_s)}{2\omega_r}\right) - \frac{V_o\sin\beta^2}{2N\omega_r Z_c}\right] \quad (5.38)$$

where, $\beta = \frac{T_r}{2\omega_r} = \frac{1}{2}T_r\sqrt{L_r C_r}$.

Fig. 5.7 Equivalent circuit of SQR for the positive half cycle

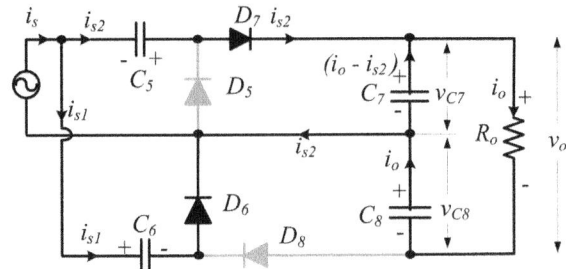

5.5.2 SQR Analysis with LLC Resonant Tank Using FHA

To show the efficacy of the proposed differential equation analysis method as compared to FHA method, it is important to derive equations based on FHA method. In this section, the equation for the output voltage, $v_o(t)$ of the converter is derived using FHA. Subsequently, the equations for the output voltage ripple, $V_{ripple,FHA}$ and the equivalent reflected load resistance, R_e are derived. It should be noted that the SQR is different from the normal rectifier due to the series (C_5, C_6) and parallel (C_7, C_8) combination of voltage multiplying capacitors.

For FHA analysis, the input current to the SQR circuit, $i_s(t)$ is assumed to have only single frequency component i.e. $i_s(t) = I_{sm} \sin \omega_s t$. As show in Fig. 5.7, the instantaneous output voltage $v_o(t)$ can be written as the sum of voltages, $v_{c7}(t)$ and $v_{c8}(t)$.

$$v_o(t) = v_{C7}(t) + v_{C8}(t) \tag{5.39}$$

Differentiating the (5.39) gives,

$$\frac{dv_o(t)}{dt} = \frac{dv_{C7}(t)}{dt} + \frac{dv_{C8}(t)}{dt} \tag{5.40}$$

Assuming $C_7 = C_8 = C_o$ and replacing $\frac{dv_{C7}(t)}{dt}$ and $\frac{dv_{C8}(t)}{dt}$ in terms of currents flowing through the capacitor C_7 and C_8 as shown in the Fig. 5.7, Eq. (5.40) can be derived as follows,

$$\frac{dv_o(t)}{dt} = \frac{-(i_o(t) - i_{s2}(t))}{C_o} + \frac{-i_o(t)}{C_o} \tag{5.41}$$

The output current, $i_o(t)$ can be replaced with $\frac{v_o(t)}{R_o}$. Similarly, the value of $i_{s2}(t)$ is simplified using (5.13) as $i_{s2}(t) = \frac{I_{sm} \alpha \sin \omega_s t + i_o(t)}{2\alpha + 1}$.

Define $k_1 = \frac{4\alpha+1}{R_o C_o (2\alpha+1)}$; $k_2 = \frac{\alpha I_{sm}}{C_o (2\alpha+1)}$. Substituting k_1 and k_2 in (5.41), the following linear differential equation is obtained.

$$\frac{dv_o(t)}{dt} + k_1 v_o(t) - k_2 \sin(\omega_s t) = 0 \tag{5.42}$$

In steady state, $v_o(0) = v_o(\pi/w_s)$.

The solution of differential equation given in (5.42) provides, $v_o(t)$,

$$v_o(t) = -\frac{k_2 \ (w_s \ \cos(w_s \ t) - k_1 \ \sin(w_s \ t))}{\left(k_1{}^2 + w_s^2\right)} - \frac{2 \ k_2 \ w_s \ e^{-k1 \ t}}{\left(k_1{}^2 + w_s^2\right) \left(e^{-\frac{\pi \ k1}{w_s}} - 1\right)} \tag{5.43}$$

The average value of $v_o(t)$ for half of the switching cycle, $< v_o(t) >$ is given by

$$< v_o(t) > = \frac{2 \, I_{sm} \, R_o \, \alpha}{\pi \, (4 \, \alpha + 1)} \tag{5.44}$$

where, R_o is the output load resistance of the converter.

The output voltage ripple based on FHA, $V_{ripple,FHA}$ is given by,

$$\begin{aligned}
V_{ripple,FHA} &= v_o\left(\frac{3\pi}{4\omega_s}\right) - v_o\left(\frac{\pi}{4\omega_s}\right) \\
&= \frac{15\sqrt{2} \, C_o \, R_o \, T_s \, V_o \, \pi^2 \left(e^{\frac{5\,T_s}{12\,C_o\,R_o}} - \sqrt{2}\,e^{\frac{5\,T_s}{24\,C_o\,R_o}} + 1\right)}{\left(e^{\frac{5\,T_s}{12\,C_o\,R_o}} + 1\right) \left(36\,\pi^2\,C_o^2\,R_o^2 + 25\,T_s{}^2\right)}
\end{aligned} \tag{5.45}$$

By equating primary side power to secondary side power and using (5.44), the equivalent reflected load resistance, R_e is derived as follows.

$$R_e = \frac{R_o}{2\pi^2 N^2(\frac{1}{4\alpha} + 1)^2} \tag{5.46}$$

The value of R_e depends on α which is a unique characteristic of the SQR based LLC converter and is not evident if approximated by the FHA analysis of a normal rectifier [22]. The derived R_e is used to define the quality factor, Q of the LLC resonant tank. In Sect. 5.5.3, another method of calculating output voltage ripple without any approximation is given. Subsequently, a comparison between the accuracy of these two results is presented in Sect. 5.7.

5.5.3 Accurate Output Voltage Ripple Estimation Using Differential Equation Based Method

The output voltage ripple can be calculated using (5.45). However, it is based on assumption that input current, i_s to the SQR is sinusoidal. Therefore, the calculation of output ripple voltage will not be accurate as the current to the SQR is discontinuous with a discontinuous time, t_d as shown in Fig. 5.4b. As stated earlier, for the application such as MPM, the estimation of accurate output voltage ripple is nec-

essary as the phase of amplified RF output depends on the output voltage and any
error in ripple estimation results in power loss when multiple MPMs are combined
for increased output RF power.

The equivalent circuit of the SQR for the positive half cycle has been shown in
Fig. 5.7. During this period, the load current i_o of constant magnitude, I_o discharges
capacitor C_7 and C_8 where as i_{s2} charges capacitor C_7. Assuming all the capacitors
of the SQR circuit to be equal ($C_{5-8} = C_o$), it can be shown that the net discharge
of output capacitor occurs when, $i_{s_2}(t) < 2I_0$ and they keep on discharging until
$i_{s_2}(t) > 2I_0$ if t_{c_1} and t_{c_2} are the time when $i_{s_2}(t) = 2I_0$, the peak to peak output
ripple, V_{ripple} can be derived as

$$V_{ripple} = \frac{1}{C_o}\left[\int_{t_{c_2}}^{T_r}(2I_0 - i_{s2}(t))\,dt + \int_{T_r}^{T_s/2}(2I_0)dt + \int_{T_s/2}^{\frac{T_s}{2}+t_{c_1}}(2I_0 - i_{s2}(t))\,dt\right] \quad (5.47)$$

Figure 5.8 shows the different operating regions of the SQR which results in output
voltage ripple, V_{ripple}. This first part of the ripple equation given in (5.47) describes
the voltage drop when i_{s2} becomes less than $2I_0$. The second part corresponds to the
voltage drop in the output capacitors during time, t_d whereas the third part of the
equation describes the voltage drop during the negative half cycle till i_{s2} becomes
equal to $2I_0$. The three parts are shown in shaded regions in Fig. 5.8 as region-
1, region-2 and region-3, respectively. It is to be noted that during time t_d, both
capacitors, C_7 and C_8 will discharge to the load and there is no charging current which
results in almost linear drop in output voltage. Increasing the switching frequency
below series resonant frequency results in larger t_d and therefore, results in additional
voltage drop and therefore, requires larger output capacitor to reduce the output
voltage ripple. The output voltage ripple equation given in (5.47) can be further
simplified by assuming the secondary diode current, i_{s2} to be symmetrical across its
peak value. With that assumption, the two instants, t_{c1} and t_{c2} when $i_{s2}(t) = 2I_0$ as
show in Fig. 5.9 can be simplified as,

$$t_{c1} = T_r - t_{c2} \quad (5.48)$$

The current $i\,s_2(t)$ can be expressed as $i\,s_2(t) = I_{s2m}\sin(\frac{\pi t}{T_r})$ where, I_{s2m} denotes the
maximum value of the current, $i_{s2}(t)$. The value of t_{c1} can be solved by equating

Fig. 5.8 Peak to peak
voltage ripple, V_{ripple} of the
output DC voltage. The
shaded regions show the
process by which output
voltage ripple is generated.
The output DC current, i_o is
of constant magnitude I_o

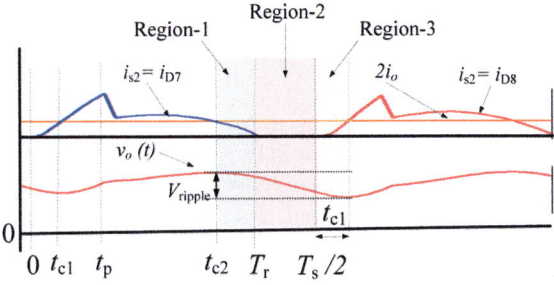

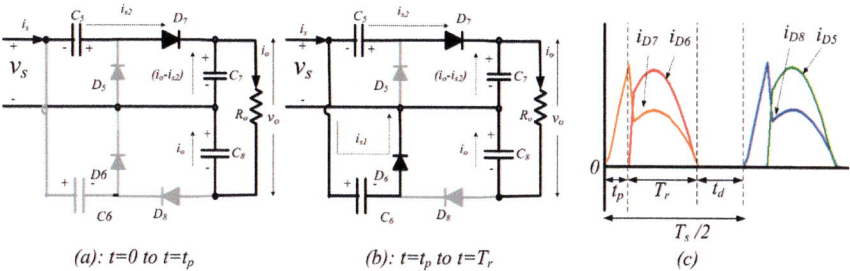

(a): $t=0$ to $t=t_p$ (b): $t=t_p$ to $t=T_r$ (c)

Fig. 5.9 **a** SQR operation during positive half cycle for $t = 0$ to $t = t_p$. **b** SQR operation during positive half cycle for $t = t_p$ to $t = T_r$. **c** Current in the SQR diodes

$i_{s2}(t) = 2I_o$. The value of t_{c1} can be derived as,

$$t_{c1} = \frac{T_r}{\pi} \sin^{-1}\left(\frac{2I_o}{I_{s2m}}\right) \tag{5.49}$$

From (5.13), (5.47), (5.48) and (5.49) the output voltage ripple can be further simplified as,

$$V_{ripple} = \frac{I_o T_s}{C_o}\left(1 - \frac{2(1 + \frac{\alpha I_{sm}}{I_o})}{\pi(2\alpha + 1)}\frac{T_r}{T_s}\right) \tag{5.50}$$

It is to be noted that the output ripple voltage is function of T_r. The decrease in T_r is characterized by increased discontinuous time period, t_d contributing to increased output ripple voltage.

5.5.4 Explanation for the Time t_p and Unequal Current Stress in the Diodes of SQR

In this section, the unique nature of the SQR diode currents is described. As shown in Fig. 5.9c the SQR diode currents, i_{D7} and i_{D6} do not start at the same time. The operation of the SQR based on the characteristic behaviour of these two current can be divided into two regions. The first region occurs for duration, $t = 0$ to $t = t_p$ as show in Fig. 5.9a whereas second region occurs for duration, $t = t_p$ to $t = T_r$ as show in Fig. 5.9b.

Since operation of the SQR is symmetrical in the both positive and negative half cycle, only positive cycle is described. Initially, the secondary current, i_s flows through diode, D_7 discharging the series capacitor, C_5 in this process which is given by the following equation,

$$v_s(t) = v_{c7}(t) - v_{c5}(t); \quad i_{s2}(t) = i_s(t); \quad i_{s1}(t) = 0; \tag{5.51}$$

Assuming $v_{c7}(t) = \frac{V_o}{2}$, the secondary voltage, $v_s(t)$ increases with the discharge of the capacitor, C_5. For the diode, D_6 to conduct, $v_s(t) - v_{c6}(t) > V_f$ where V_f is the forward voltage drop of the diode, D_6. The time, t_p is essentially the duration when the secondary voltage becomes higher than the voltage across the capacitor, C_6 and therefore, both diodes, D_6 and D_7 start conducting as shown in Fig. 5.9b. As the impedance offered by two paths, C_5 - D_7 - C_7 and C_6 - D_6 are not equal, the currents shared by the two diodes, D_7 and D_6 are not equal as described by the Eqs. (5.12) and (5.13). However, increasing the value of capacitor, C_7 in relative to C_5 and C_6, the difference in the impedance of the two paths two paths, C_5 - D_7 - C_7 and C_6 - D_6 can be minimized. As, the parameter α describes the ratio of the parallel capacitors (C_7, C_8) and the series capacitors, (C_5, C_6), the increase in the value of α reduces the difference of the currents in the diodes, D_7 and D_6.

5.5.5   Effect of α on the Operation of the SQR

By choosing α more than 1, the difference of two currents i_{s1} and i_{s2} can be reduced as given by (5.12) and (5.13). Moreover, the time interval, t_p also gets reduced with increased value of α. However, large α requires small series capacitors, C_5 and C_6. The small capacitance results in increased voltage ripple for a current fed SQR circuit. For high voltage output, the current in secondary side is relatively smaller and therefore, the value of α should not be chosen bigger as it increases the voltage stress of the capacitors by increasing the voltage ripple across them. However, for the converters having low voltage and high current output, large value of α can effectively reduce the current stress of the SQR diodes.

5.5.6   Current Stress in SQR Diodes

From (5.44), the ratio i_{sm} and I_o which is a function of α is calculated. Figure 5.10 shows the variation of the secondary peak current, I_{sm} with respect to α. With increase in α, the peak current of the transformer secondary current, i_{sm} is reduced. Subsequently, the SQR diode currents, i_{s1} and i_{s2} can be calculated from (5.12) and (5.13). Figure 5.11 shows the variation of the peak of the SQR diode current with respect to α. With increase in α, the peak current stress of the SQR diodes tends to be equal.

5.5.7   Voltage Stress in the Capacitors of the SQR

In the steady state the voltage across the parallel capacitors, C_7 and C_8 is half of the output voltage whereas the voltage across the series capacitors, C_5, C_6 is one-fourth

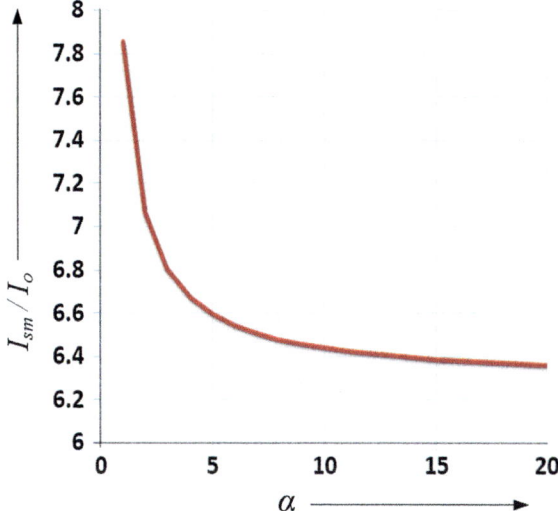

Fig. 5.10 Ratio of the peak value of the transformer secondary current and output DC current, $\frac{I_{sm}}{I_o}$ variation with respect to α

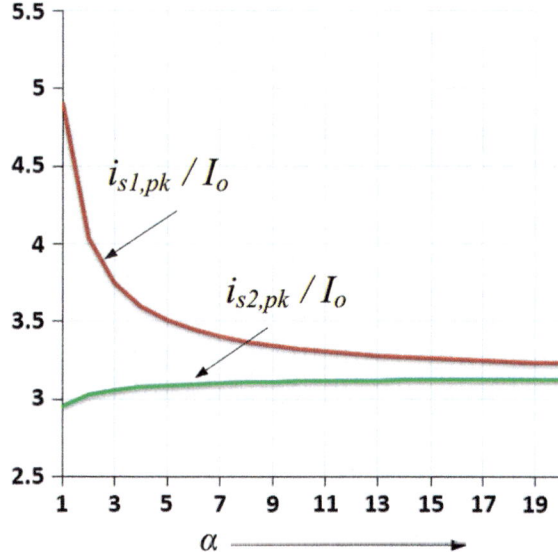

Fig. 5.11 Peak current stresses of the SQR diodes with respect to α

of the output voltage. However, all the capacitors have additional voltage ripple which must be taken into consideration for deciding the voltage rating of each of the capacitors.

The desired output voltage ripple decides the lower limit of the parallel capacitors (5.47) whereas its upper limit is constrained by maximum allowed stored energy (E_{max}) in the output side.

$$E_{max} \leq \frac{1}{2}CV_o^2 \tag{5.52}$$

Once the parallel capacitors, C_7, C_8 are selected, the series capacitors, C_5, C_6 are chosen. For increased value of α, the series capacitors should be of lower value compared to the parallel capacitors. However, lower value of series capacitor results in increased output voltage ripple. Therefore, a compromise needs to be made between reducing the current stress (by increasing the α) and reducing the voltage stress (by keeping $\alpha = 1$). For high voltage DC output, voltage stress is much of importance than current stress. Therefore, $\alpha = 1$ is chosen in the design.

5.6   Design of the SQR Based High Voltage DC–DC Converter

In this section, comprehensive design of the presented converter is explained. A design flow chart is provided to show step by step design procedure of the converter. Subsequently, the various design parameters of the converter for a given specifications are evaluated which are later used in evaluating the performance of the converter through simulation and experimental tests. Moreover, the design of SQR circuit is also provided. An example converter of the specification shown in Table 5.1 is chosen for design. The design flow chart shown in Fig. 5.12 is used to carry out the design of the converter. Based on the design steps, for the input voltage, $V_{DC} = 120\,\text{V}$, output voltage, $V_o = 2000\,\text{V}$, the transformer turn ratio is selected to be 1 : 4. For a turns ratio of 1 : 4, the additional voltage gain required from the LLC resonant tank is 1.04. The resonant frequency, f_r of the converter is chosen to be 270 kHz which is sufficient enough to allow the switching frequency, f_s variation for providing additional voltage gain at different operating conditions. The quality factor of the LLC resonant tank, Q is defined as,

$$Q = \frac{1}{R_e}\sqrt{\frac{L_r}{C_r}} = \frac{Z_c}{R_e} \tag{5.53}$$

where, R_e is equivalent reflected resistance in the primary side given by (5.46).

Once the LLC tank parameters are defined, the unknown parameters of the equations describing the operation of the presented converter are evaluated by solving the simultaneous equations shown in (5.34)–(5.37). The results of the solution are plotted in Figs. 5.13, 5.14 and 5.15. For given input voltage, v_{DC}, output voltage, V_o,

Table 5.1   Specification of the prototype converter

Parameters	Values
Input voltage, V_{in}	120 V
Output voltage, V_o	2000 V
Output Power, P_o	200 W

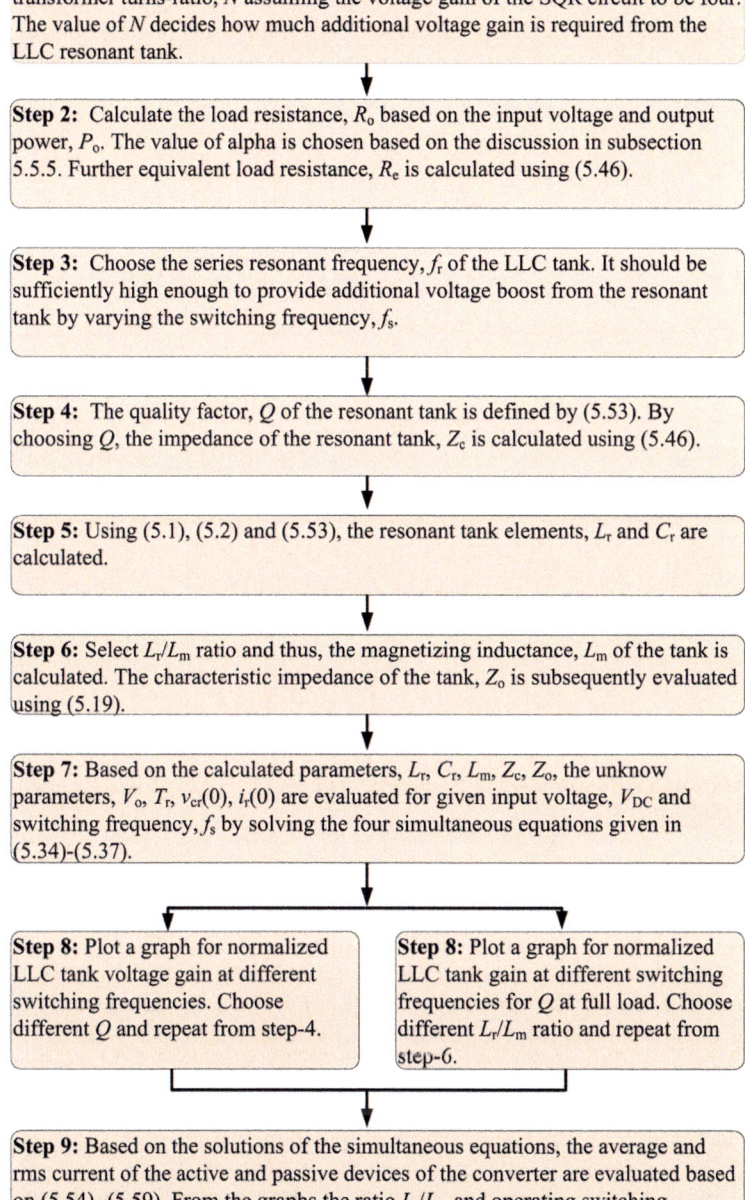

Step 1: Based on the output voltage, V_o and input voltage, V_{DC}, choose the transformer turns-ratio, N assuming the voltage gain of the SQR circuit to be four. The value of N decides how much additional voltage gain is required from the LLC resonant tank.

Step 2: Calculate the load resistance, R_o based on the input voltage and output power, P_o. The value of alpha is chosen based on the discussion in subsection 5.5.5. Further equivalent load resistance, R_e is calculated using (5.46).

Step 3: Choose the series resonant frequency, f_r of the LLC tank. It should be sufficiently high enough to provide additional voltage boost from the resonant tank by varying the switching frequency, f_s.

Step 4: The quality factor, Q of the resonant tank is defined by (5.53). By choosing Q, the impedance of the resonant tank, Z_e is calculated using (5.46).

Step 5: Using (5.1), (5.2) and (5.53), the resonant tank elements, L_r and C_r are calculated.

Step 6: Select L_r/L_m ratio and thus, the magnetizing inductance, L_m of the tank is calculated. The characteristic impedance of the tank, Z_o is subsequently evaluated using (5.19).

Step 7: Based on the calculated parameters, L_r, C_r, L_m, Z_c, Z_o, the unknow parameters, V_o, T_r, $v_{cr}(0)$, $i_r(0)$ are evaluated for given input voltage, V_{DC} and switching frequency, f_s by solving the four simultaneous equations given in (5.34)-(5.37).

Step 8: Plot a graph for normalized LLC tank voltage gain at different switching frequencies. Choose different Q and repeat from step-4.

Step 8: Plot a graph for normalized LLC tank gain at different switching frequencies for Q at full load. Choose different L_r/L_m ratio and repeat from step-6.

Step 9: Based on the solutions of the simultaneous equations, the average and rms current of the active and passive devices of the converter are evaluated based on (5.54)- (5.59). From the graphs the ratio L_r/L_m and operating switching frequency, f_s are determined for required voltage gain.

Fig. 5.12 Flowchart for the design of the presented converter

output power, P_o and the chosen converter parameters, L_r, L_m and C_r, the unknown parameters such as, $i_r(0)$, $i_m(0)$, $v_{cr}(0)$, T_r and T_s can be calculated and thus, the Eqs. (5.28), (5.29), (5.30), (5.32), (5.33) and (5.37) can be completely characterized. Subsequently, the rms, avg and peak currents of the converter are calculated using following equations.

$$i_r(avg) = \frac{2}{T_s}\left(\int_0^{T_r} i_r(t)dt + \int_{T_r}^{\frac{T_s}{2}} i_r(t)dt\right) \tag{5.54}$$

$$i_r(rms) = \sqrt{\frac{2}{T_s}\left(\int_0^{T_r} i_r^2(t)dt + \int_{T_r}^{\frac{T_s}{2}} i_r^2(t)dt\right)} \tag{5.55}$$

$$i_p(avg) = \frac{2}{T_s}\left(\int_o^{T_r} (i_r(t) - i_m(t))dt\right) \tag{5.56}$$

$$i_p(rms) = \sqrt{\frac{2}{T_s}\left(\int_o^{T_r} (i_r^2(t) - i_m^2(t))dt\right)} \tag{5.57}$$

As shown in Fig. 5.4, the peak value of resonant current, $i_r(t)$ occurs at $t = t_3$. The time t_3 is calculated by equating $i_m(t) = 0$.

$$t_3 = 4NL_m\left(\frac{-i_m(0)}{V_o}\right) \tag{5.58}$$

The peak value of the resonant current, $i_r(t)$ and the primary current, $i_p(t)$ are equal and can be given as,

$$i_r(pk) = i_p(pk) = Ni_{sm} = i_r(t_3) \tag{5.59}$$

To operate the LLC resonant converter in the boost mode, it must be operated at f_s lower than its series resonant frequency, f_r. However, f_s should also be above the second resonant frequency, f_0 in all operating conditions so that the converter can operate in ZVS mode. The characteristic plot of the converter is shown in Fig. 5.13. It shows the normalized voltage gain of the LLC tank at different $\frac{f_s}{f_r}$ for different Q values. The $\frac{L_r}{L_m}$ ratio is chosen as 0.07. It is evident from the plot that the voltage gain of the LLC tank is more than 1. Decreasing the load of the converter reduces the quality factor, Q of the LLC tank which results in increased voltage gain as shown in Fig. 5.13.

The selection of $\frac{L_r}{L_m}$ ratio is critical as it decides the voltage gain, RMS switch current and peak of the magnetizing current. In Fig. 5.14, the normalized voltage gain of the LLC tank is plotted at different $\frac{f_s}{f_r}$ for different values of $\frac{L_r}{L_m}$. For higher values of $\frac{L_r}{L_m}$ ratio, the LLC tank provides higher voltage gain. However, the higher values of $\frac{L_r}{L_m}$ also results in higher RMS current in MOSFET switches resulting in

Fig. 5.13 Normalized voltage gain of the LLC tank for different $\frac{f_s}{f_r}$ ratio at different Q

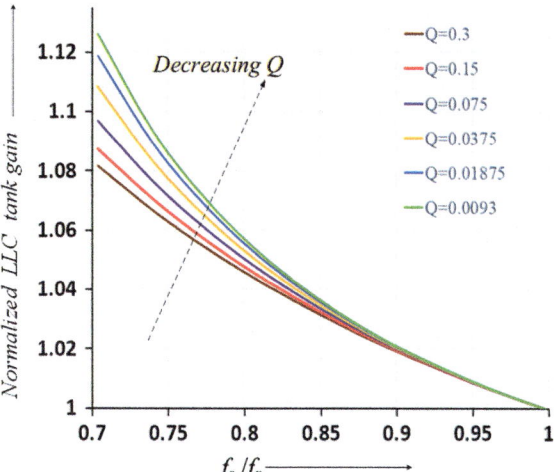

Fig. 5.14 Normalized voltage gain of the LLC tank for different $\frac{f_s}{f_r}$ ratio at different $\frac{L_r}{L_m}$ ratio

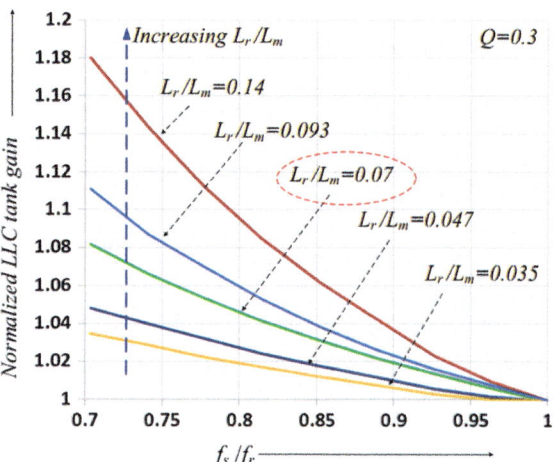

higher conduction loss as shown in Fig. 5.15a. Similarly, for lower $\frac{L_r}{L_m}$ values, the voltage gain of LLC tank decreases and therefore, the switching frequency, f_s of the converter has to be decreased further for increased voltage gain as shown in Fig. 5.15b. However, at lower $\frac{L_r}{L_m}$ values, the peak of magnetizing current reduces which is essentially turn off current for the MOSFETs resulting in reduced switching loss as illustrated in Fig. 5.15a. Therefore, a trade off is required while selecting $\frac{L_r}{L_m}$ ratio. The value of $\frac{L_r}{L_m}$ is chosen to be 0.07 as the switch RMS current is almost equal to the condition when $\frac{L_r}{L_m} = 0.035$ as shown in Fig. 5.15a. Moreover, the switching frequency, f_s is 230 kHz for required voltage gain. The switch RMS current and switch turn off current can be reduced by reducing $\frac{L_r}{L_m}$ ratio. However, it results in

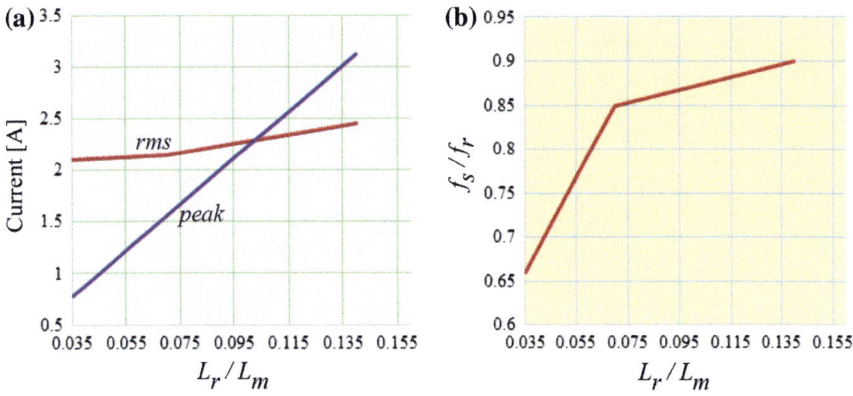

Fig. 5.15 **a** RMS and peak value of the series resonant current, i_r variation with respect to $\frac{L_r}{L_m}$. **b** $\frac{f_s}{f_r}$ variation with respect to $\frac{L_r}{L_m}$. All the graphs are plotted for full load $Q = 0.3$

lower voltage gain and lower operating switching frequency i.e. 178 kHz for $\frac{L_r}{L_m} = 0.035$. Therefore, $\frac{L_r}{L_m}$ ratio is chosen 0.07 for the presented converter.

The design of capacitors of the SQR is based on the specified output voltage ripple. However, the maximum value of capacitance is limited by maximum allowable stored energy in the output side. For 0.5 J of output energy and 2 kV output voltage,

$$\frac{1}{2}C_7\left(\frac{V_o}{2}\right)^2 + \frac{1}{2}C_8\left(\frac{V_o}{2}\right)^2 \le 0.5 \tag{5.60}$$

which gives, $C_5 = C_6 \le 0.5\,\mu\text{F}$

The output voltage ripple is calculated based on (5.47). The voltage ripple across series capacitors, C_5 and C_6 from (5.61). The SQR diodes is chosen based on average current rating given in (5.62). The voltage rating of diodes should be more than $\frac{V_o}{2}$.

$$\Delta V_{C6} = \frac{1}{C_6}\int_0^{T_r} i_{s1}(t)dt;\ \Delta V_{C5} = \frac{1}{C_5}\int_0^{T_r} i_{s1}(t)dt \tag{5.61}$$

$$< i_{D5} >=< i_{D6} >= \frac{1}{T_s}\int_0^{T_r} i_{s1}(t)dt;$$

$$< i_{D7} >=< i_{D8} >= \frac{1}{T_s}\int_0^{T_r} i_{s2}(t)dt \tag{5.62}$$

For switching frequency, f_s is 230 kHz and $\frac{L_r}{L_m} = 0.07$, the different parameters of converter are designed and are tabulated in Table 5.2.

For the designed parameter shown in Table 5.2, the RMS and *avg* currents of the active and passive devices are evaluated for full load and 50% of the full load by solving the design equation given in (5.34)–(5.37). The results are tabulated in Table 5.3.

Table 5.2 Designed parameters of the prototype converter

Parameter	Value
Series resonant inductor, L_r	$7\,\mu H$
Resonant capacitor, C_r	$50\,nF$
Magnetizing inductance, L_m	$100\,\mu H$
Transformer turns-ratio, $1:N$	$1:4$
SQR capacitor, C_0	$75\,nF$

Table 5.3 Theoretical calculation of currents by solving the iterative design equations for 200 and 100 W output power

P_o	200 W	100 W
$i_{r,rms}$	2.00 A	1.27 A
$i_{r,avg}$	1.80 A	1.16 A
$i_{r,pk}$	2.96 A	1.51 A
$i_{p,rms}$	1.93 A	0.97 A
$i_{p,avg}$	1.61 A	0.8 A
$i_{p,pk}$	2.96 A	1.51 A
$i_{s,rms}$	0.48 A	0.24 A
$i_{s,avg}$	0.4 A	0.20 A
$i_{s,pk}$	0.74 A	0.38 A

5.7 Simulation and Experimental Results

A scaled down hardware prototype of 200 W and 2000 V is designed as shown in Fig. 5.18. The designed parameters of the converter is tabulated in Table 5.2. To generate the control signal for the full bridge converter, TI F28335 DSP board is used. Power MOSFETs (IRFI4227PBF) are used to implement the full bridge semiconductor switches. To isolate the control signal and MOSFETs, pulse transformer based gate drivers are used. The high frequency AC output of the full bridge inverter is followed by a resonant tank and a high frequency transformer. To have low profile and sufficient voltage isolation, a high frequency planar transformer with $1:4$ turns ratio is designed with ferrite core, EE-32-3C96 as shown in Fig. 5.19. The leakage inductance of the transformer is found to be $3.3\,\mu H$. A magnetizing inductance of $100\,\mu H$ is designed by creating an air-gap between the two E cores. To form the LLC resonant tank an external inductor of $4\,\mu H$ and an external capacitor of $50\,nF$ is connected in series. To reduce the voltage and current stress in the SQR diodes / capacitors, the secondary winding is divided into two similar sections. The SQR capacitor value in each of the sections is chosen to be $0.075\,\mu F$. The rectified voltage of these two sections is added in series to obtain the total output voltage. Ultrafast 600 V(MUR860G) diodes are used in the SQR circuit.

5.7.1   Simulation Results

The presented converter with above specifications is simulated in PSIM 9.3 and the simulation results are provided. The parameters of the converter for simulation are chosen as shown in Table 5.2. A variable frequency controller for regulating the output voltage is implemented in simulation and results are obtained for 120 VDC input, 2000 VDC output at 200 W output power. The converter is operated at 230 kHz switching frequency.

Figure 5.16 shows the simulation results for the converter at full load (200 W). Figure 5.16a shows the full bridge inverter voltage, v_{in}. In Fig. 5.16b, the series resonant current, i_r and the magnetizing current, i_m are shown. It is evident that LLC tank is operating at switching frequency lower than the series resonant frequency. Figure 5.16c shows the secondary current, i_s of the transformer which is also input to the SQR circuit. As shown, the current is discontinuous in nature. Figure 5.16d shows the diode currents of the SQR circuit whereas Fig. 5.16e shows the regulated DC output voltage, v_o.

In Fig. 5.17, the series resonant current, i_r and the magnetizing current, i_m are shown for three different values of $\frac{L_r}{L_m}$. As shown, with increase in $\frac{L_r}{L_m}$ value from 0.035 to 0.14, the switching frequency, f_s of the converter is increasing. Moreover, the peak of magnetizing current, i_m is maximum at $\frac{L_r}{L_m} = 0.14$ as shown in Fig. 5.17a. The simulation results shown in Fig. 5.17 validates the results shown in Fig. 5.15 and justifies the choice of $\frac{L_r}{L_m} = 0.07$ in the design.

5.7.2   Experimental Results

The hardware prototype of the converter shown in Fig. 5.18 is tested for 2 kV output and experiment results are discussed. A ferrite core based high frequency planar transformer is designed for step up voltage gain as shown in Fig. 5.19. Figure 5.20 shows the full bridge inverter voltage v_{in} and series resonant current, i_r for 200 W output power. The input voltage is 120 VDC and output voltage is 2000 VDC. The switching frequency of the converter is 230 kHz. The input current of the resonant tank, i_r is made of two components- primary current, i_p and magnetizing current, i_m. It is evident from the waveform of i_r that the converter is operating below series resonant frequency, f_r. Figure 5.21 shows switch voltage (V_{ds}), gate voltage (V_{gs}) and series resonant current, i_r for 120/2000 V at 200 W output power. It can be noticed that V_{ds} becomes almost zero before the gate is turned ON ($V_{gs} > 0$). The series resonant current discharges the switch capacitance before it is turned ON and therefore, guarantees ZVS for the switches. Similarly, Fig. 5.22 shows the switch voltage (V_{ds}), gate voltage (V_{gs}), series resonant current (i_r) and output voltage (V_0) at 50% of the full load. It can be seen from the waveforms that the switch is turned ON with zero voltage in this condition too.

Fig. 5.16 Simulation Results of the presented converter. **a** Full bridge inverter output voltage, v_{in}, **b** Series resonant current, i_r and magnetizing current, i_m, **c** Secondary transformer current, i_s, **d** Diode currents of the SQR, **e** High voltage DC output, V_o

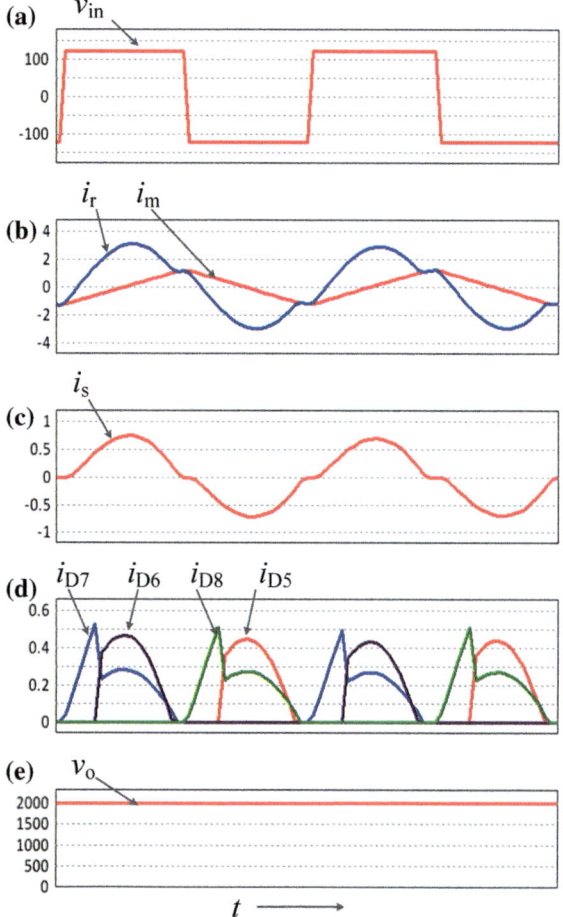

Figure 5.23 shows transformer secondary voltage (v_s) and secondary current (i_s) at 200 W output power. It is to be noted here that the secondary windings of transformer is divided into two similar sections and the voltage and current waveforms are taken for one of those sections. The secondary current is discontinuous indicating the operation of converter below f_r. As i_s becomes zero for t_d duration after each half of the switching cycle, the diodes of SQR turn off with zero current switching (ZCS) contributing to improved power conversion efficiency.

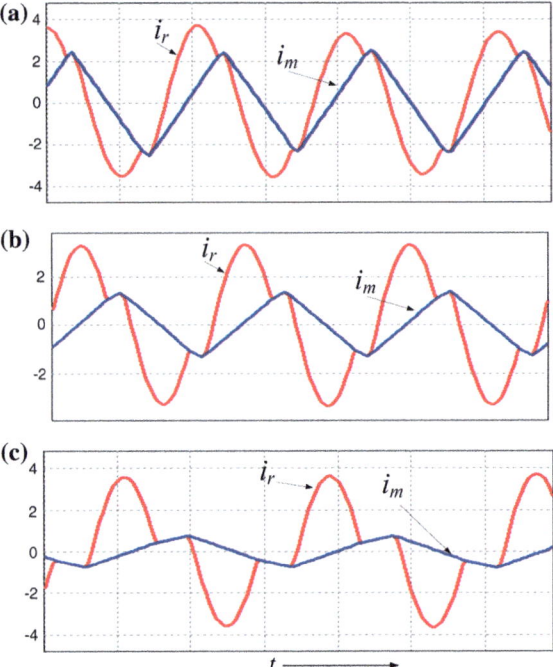

Fig. 5.17 Simulated Series Resonant current, i_r and magnetizing current, i_m for different $\frac{L_r}{L_m}$ ratio **a** $\frac{L_r}{L_m} = 0.14$, **b** $\frac{L_r}{L_m} = 0.07$, **c** $\frac{L_r}{L_m} = 0.035$

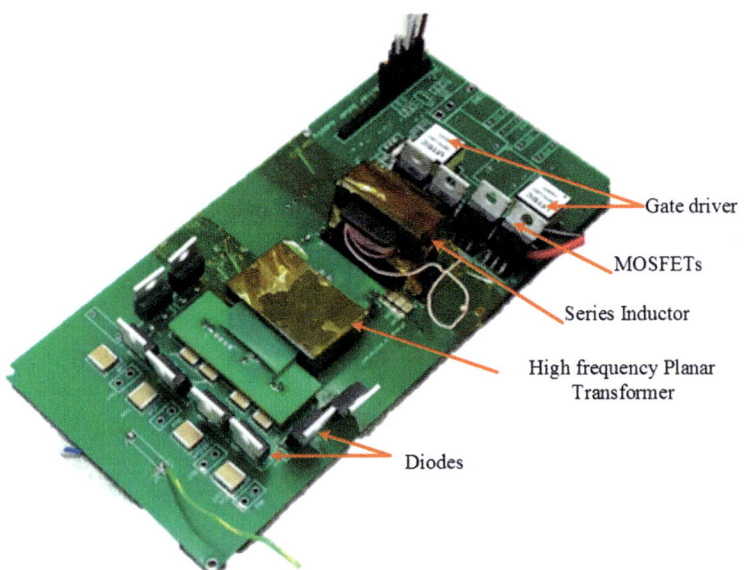

Fig. 5.18 Hardware prototype of the proposed high voltage DC–DC converter

Fig. 5.19 Hardware prototype of the high frequency step up transformer

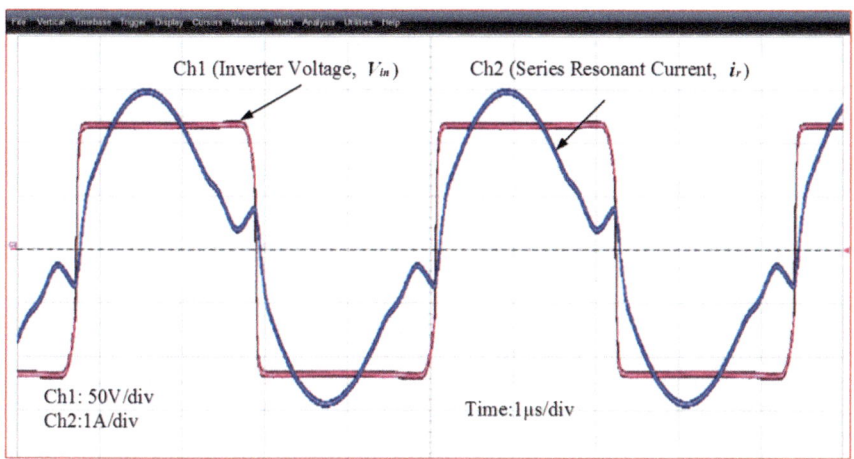

Fig. 5.20 Full bridge inverter voltage, (v_{in}) and series resonant current, (i_r) for 200 W output power at 120 V DC input to 2 kV output. Switching frequency (f_s) is 230 kHz

Figure 5.24 shows the high voltage DC output ($V_0 = 2000$ V) at 200 W output power. The input voltage is 120 V and $f_s = 230$ kHz. Figure 5.25 shows the output voltage ripple (V_{ripple}) at 200 W output power. The ripple frequency is two times (460 kHz) of the switching frequency, f_s. The theoretical ripple voltage is calculated using (5.50) for 200 and 100 W output power and are found to be 4.2 and 1.9 V. Figure 5.26 shows the output voltage ripple (V_{ripple}) at 100 W output power. The ripple voltage for 200 W is found to be 4 V where as for 100 W, it is 2.1 V.

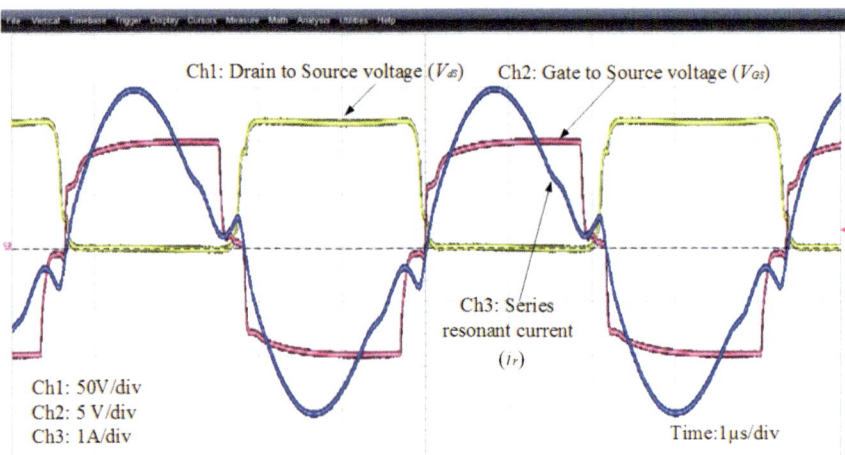

Fig. 5.21 Switch voltage (V_{ds}), Gate voltage (V_{gs}), and series resonant current (i_r) for 120 V DC/2000 V at 200 W output power

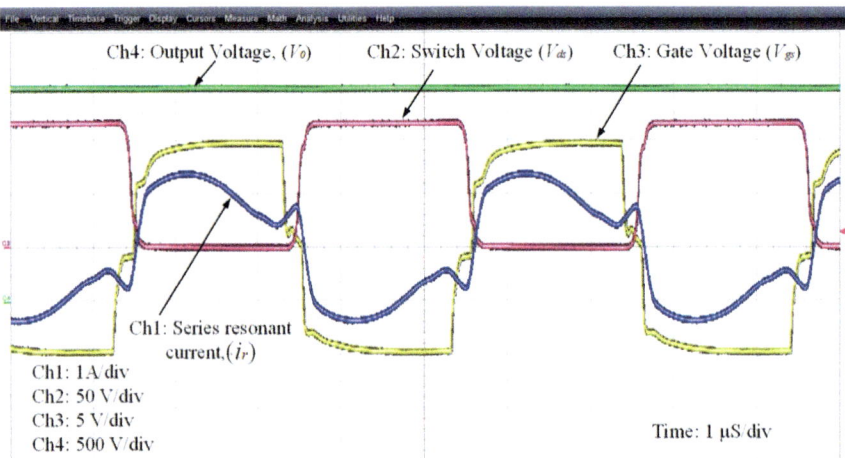

Fig. 5.22 Switch voltage (V_{ds}), Gate voltage (V_{gs}), high voltage DC output (V_0) and series resonant current (i_r) for 120 V DC/ 2 kV at 100 W output power

5.7.3 Discussion of the Power Conversion Efficiency of the Converter

The theoretical power loss of the presented converter is calculated and the distribution of the power losses for different parts of the converter are shown for operation at the full load. Based on calculation in Table 5.3, the RMS current in the switches at full load is 2.09 A. As the converter is having Zero Voltage Switching (ZVS) during

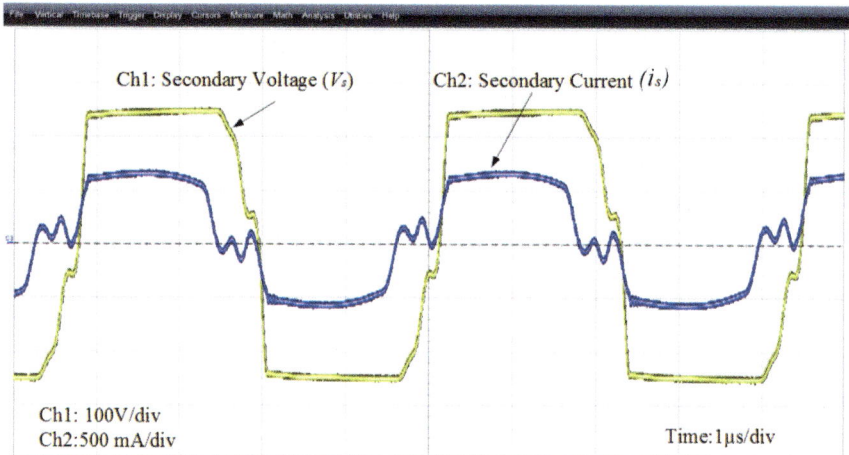

Fig. 5.23 Transformer secondary voltage, v_s and current waveforms, i_s at 200 W output power and 120 V DC input

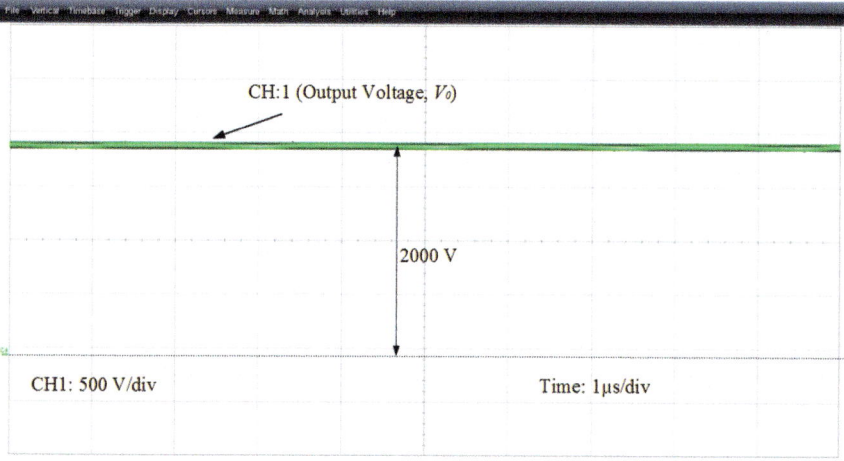

Fig. 5.24 High voltage DC output (V_0). The presented DC–DC converter generates 2000 V from 120 V DC input when converter is switching at 230 kHz at 200 W output power

turn ON, the switch turn ON losses are assumed to be zero. However, during turn off, the MOSFET generates loss as it turns off with the peak of magnetizing current, i_m which value depends on $\frac{L_r}{L_m}$ ratio. For $\frac{L_r}{L_m} = 0.07$, the peak of the magnetizing current is 1.32A and therefore, the turn off losses of the switches are considered into loss calculation. Further, loss in external inductor is calculated. The core loss and copper loss of the high frequency transformer is calculated based on [23]. The loss in resonant capacitor, C_r is calculated based on its ESR values. In the secondary side, the conduction loss in the SQR diodes are calculated based on the average current

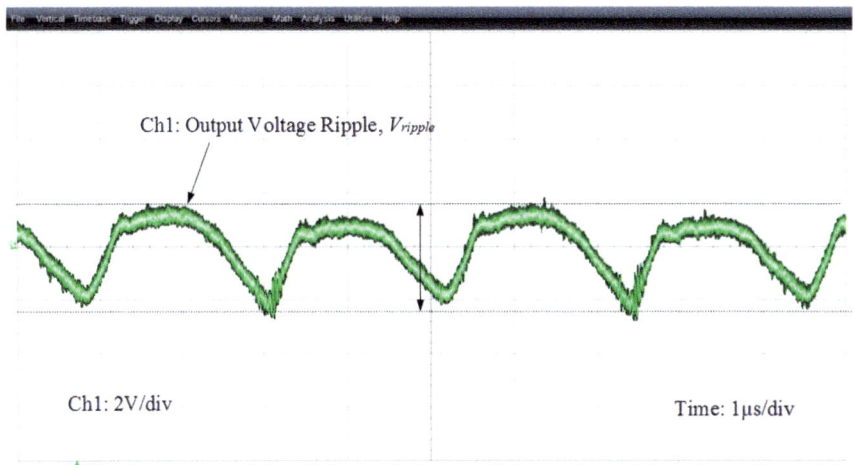

Fig. 5.25 Output voltage ripple (V_{ripple}) of high voltage DC output at 200 W output power and 120 V DC input

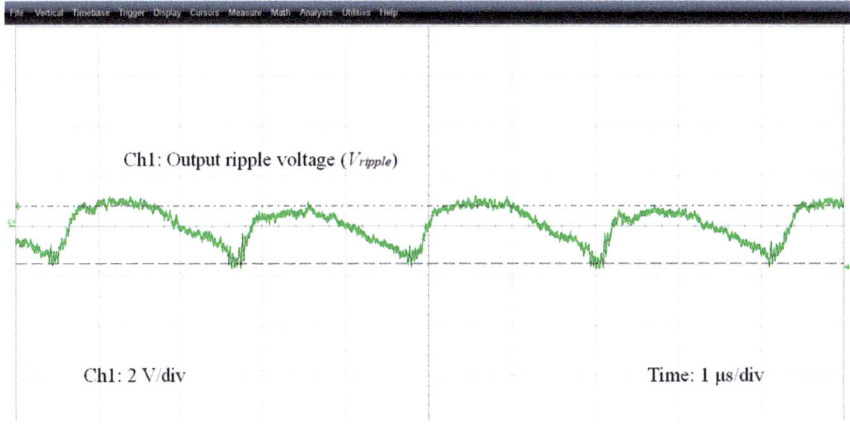

Fig. 5.26 Output voltage ripple (V_{ripple}) of high voltage DC output at 100 W output power and 120 V DC input

flowing through them which is calculated and tabulated in Table 5.3. Since the SQR diodes switch with zero current, the reverse recovery losses in the SQR diodes are zero. The losses due to ESR of the SQR capacitors are also considered. The total loss of the converter is summarized in Table 5.4 and its distribution is shown in Fig. 5.27.

The losses in the MOSFETs include conduction loss and turn off loss. The MOSFET used in implementing the hardware prototype, IRFI4227PBF offers very low $R_{ds,on}$ (23 mΩ) which results in lower conduction loss (86 mW per MOSFET). However, as turn off current in the presented converter is high due to low magnetizing inductance, the turn off losses are significant (720 mW per MOSFET). The total loss

Table 5.4 Loss Distribution in the different parts of the converter

Gate drive loss	0.952 W
Mosfet loss	3.26 W
Resonant inductor loss	0.107 W
Resonant capacitor loss	0.441 W
Transformer loss	2.2 W
SQR loss	0.613 W
Total loss	7.57 W

Fig. 5.27 Theoretical power loss distribution in watts for the presented converter. For 200 W output power, the total loss is 7.57 W

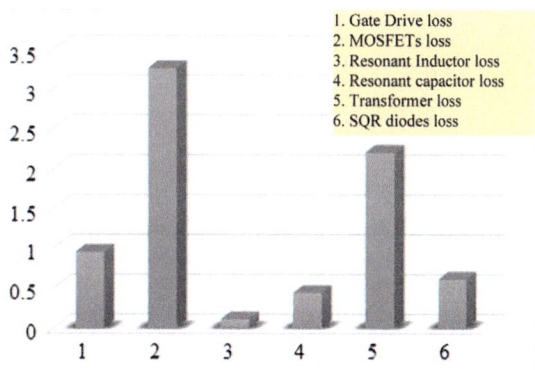

1. Gate Drive loss
2. MOSFETs loss
3. Resonant Inductor loss
4. Resonant capacitor loss
5. Transformer loss
6. SQR diodes loss

of the MOSFETs can be further improved by the proper selection of MOSFETs. However, selection of the optimum switching device has not been focused in this research work. Similarly, the transformer loss includes core loss and copper loss. The operating magnetic flux density is chosen to be 0.1 T for the transformer. The core loss and copper loss are 1.18 and 1.021 W, respectively. The SQR loss is the sum of the losses in diodes and capacitors. The losses in diodes and capacitors are 500 and 109 mW respectively. The total theoretical loss is 7.57 W which is very close to experimental loss of 8 W at 200 W output power. The overall efficiency of the converter at full load is found to be 96%. The efficiency of the converter is tested at 50% of the full load and an experimental efficiency of 94.1% is achieved.

5.7.4 Comparison Between Proposed Differential Equation Based Analysis with FHA Method

The first harmonic approximation simplifies the analysis of an LLC resonant converter. However, it fails to provide accurate results when the converter is operated at switching frequency lower than the series resonant frequency to have additional voltage gain from the LLC tank. The proposed differential equation based analy-

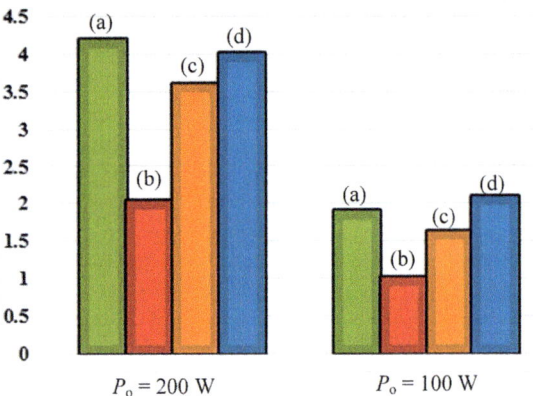

Fig. 5.28 Comparison of the accuracy in estimating the output voltage ripple of the presented converter by using the FHA and the proposed method. **a** Proposed method, **b** FHA method, **c** Digital simulation results, **d** Experimental results

sis provides accurate results as it does not take any approximation into account. To demonstrate the efficacy of the proposed method, the output voltage ripple is calculated using both the FHA and the proposed differential equation based method by (5.45) and (5.47), respectively. Subsequently, the results are compared with simulation and experimental results. Figure 5.28 shows the comparison of the accuracy in estimating the output voltage ripple of the converter using the FHA and the proposed method.

It is evident that the FHA method fails to provide accurate output voltage ripple. The proposed analysis method provides results much closer to simulation and experimental results both at 200 and 100 W output power.

5.8 Conclusion

A 200 W, 120 V/2 kV symmetrical quadrupler rectifier based LLC resonant converter with high voltage gain has been presented in this chapter along with comprehensive mathematical analysis and extensive experimental results to validate the feasibility of the presented converter. The LLC converter is operated at lower than the series resonant frequency of the LLC tank to have additional voltage boost. During this mode, the primary current in the tank becomes discontinuous and therefore, the usual design based on the FHA does not hold true. Therefore, a new method of solving the LLC resonant tank without any approximation is proposed. Subsequently, the efficacy of the proposed method is validated through simulation and experimental results. The output voltage ripple estimation through both FHA and the proposed method show that the proposed method accurately estimate the ripple with just 5% error whereas as the FHA method shows significant error of 50% when compared with the experimental results. The use of SQR along with an additional voltage gain from the LLC tank reduces the required turns ratio of the transformer almost by the factor of 4 contributing to high frequency operation and reduced transformer size. An experimental efficiency of 96% is achieved at full load.

References

1. R. Casanueva, C. Brañas, F.J. Azcondo, F.J. Diaz, Teaching resonant converters: properties and applications for variable loads. IEEE Trans. Ind. Electron. **57**(10), 3355–3363 (2010)
2. B.S. Nathan, V. Ramanarayanan, Analysis, simulation and design of series resonant converter for high voltage applications, in *Proceedings of IEEE International Conference on Industrial Technology 2000*, vol. 1 (IEEE, 2000), pp. 688–693
3. A. Santoja, A. Barrado, C. Fernandez, M. Sanz, C. Raga, A. Lazaro, High voltage gain dc-dc converter for micro and nanosatellite electric thrusters, in *Applied Power Electronics Conference and Exposition (APEC), 2013 Twenty-Eighth Annual IEEE*, (IEEE, 2013), pp. 2057–2063
4. J. Liu, L. Sheng, J. Shi, Z. Zhang, X. He, Design of high voltage, high power and high frequency transformer in lcc resonant converter, in *Applied Power Electronics Conference and Exposition, 2009. APEC 2009. Twenty-Fourth Annual IEEE*, (IEEE, 2009) pp. 1034–1038
5. C.-H. Chang, E.-C. Chang, H.-L. Cheng, A high-efficiency solar array simulator implemented by an llc resonant dc-dc converter. IEEE Trans. Power Electron. **28**(6), 3039–3046 (2013)
6. F.G. Turnbull, R.E. Tompkins, Design of a pulsewidth-modulated resonant converter for a high-output-voltage power supply. IEEE Trans. Ind. Appl. **6**, 1016–1021 (1987)
7. S.-R. Jang, H.-J. Ryoo, S.-H. Ahn, J. Kim, G.H. Rim, Development and optimization of high-voltage power supply system for industrial magnetron. IEEE Trans. Ind. Electron. **59**(3), 1453–1461 (2012)
8. S.S. Lee, S. Iqbal, M. Kamarol, Control of zcs-sr inverter-fed voltage multiplier-based high-voltage dc-dc converter by digitally tuning tank capacitance and slightly varying pulse frequency. IEEE Trans. Power Electron. **27**(3), 1076–1083 (2012)
9. J.P.V. Cunha, M. Begalli, M.D. Bellar, High voltage power supply with high output current and low power consumption for photomultiplier tubes. IEEE Trans. Nucl. Sci. **59**(2), 281–288 (2012)
10. C.-M. Young, M.-H. Chen, T.-A. Chang, C.-C. Ko, K.-K. Jen, Cascade cockcroft-walton voltage multiplier applied to transformerless high step-up dc-dc converter. IEEE Trans. Ind. Electron. **60**(2), 523–537 (2013)
11. S. Iqbal, A three-phase symmetrical multistage voltage multiplier. IEEE Power Electron. Lett. **3**(1), 30–33 (2005)
12. S. Iqbal, A hybrid symmetrical voltage multiplier. IEEE Trans. Power Electron. **29**(1), 6–12 (2014)
13. T.B. Soeiro, J. Muhlethaler, J. Linner, P. Ranstad, J.W. Kolar, Automated design of a high-power high-frequency lcc resonant converter for electrostatic precipitators. IEEE Trans. Ind. Electron. **60**(11), 4805–4819 (2013)
14. T. Guo, C. Zhang, L. Chang, J. Liu, J. Du, X. He, Large-signal modeling of lcc resonant converter operating in discontinuous current mode applied to electrostatic precipitators, in *Applied Power Electronics Conference and Exposition (APEC), 2013 Twenty-Eighth Annual IEEE*, (IEEE, 2013), pp. 2629–2635
15. S. Gavin, M. Carpita, P. Ecoeur, H.-P. Biner, M. Paolone, E.T. Louokdom, A digitally controlled 125 kvdc, 30kw power supply with an lcc resonant converter working at variable dc-link voltage: full scale prototype test results (2014)
16. J. Martin-Ramos, A. Pernia, J. Diaz, F. Nuno, J. Martinez, Power supply for a high-voltage application. IEEE Trans. Power Electron. **23**, 1608–1619 (2008)
17. M. Youssef, H. Pinheiro, P. Jain, Self-sustained phase-shift modulated resonant converters: modeling, design, and performance. IEEE Trans. Power Electron. **21**, 401–414 (2006)

18. Y. Zhao, X. Xiang, W. Li, X. He, C. Xia, Advanced symmetrical voltage quadrupler rectifiers for high step-up and high output-voltage converters. IEEE Trans. Power Electron. **28**(4), 1622–1631 (2013)
19. F. Musavi, M. Craciun, D.S. Gautam, W. Eberle, W.G. Dunford, An llc resonant dc-dc converter for wide output voltage range battery charging applications. IEEE Trans. Power Electron. **28**(12), 5437–5445 (2013)
20. D. Fu, Y. Liu, F.C. Lee, M. Xu, A novel driving scheme for synchronous rectifiers in llc resonant converters. IEEE Trans. Power Electron. **24**(5), 1321–1329 (2009)
21. D. Huang, S. Ji, F. Lee et al., Llc resonant converter with matrix transformer, in *Applied Power Electronics Conference and Exposition (APEC), 2014 Twenty-Ninth Annual IEEE*, (IEEE, 2014), pp. 1118–1125
22. G. Ivensky, S. Bronshtein, A. Abramovitz, Approximate analysis of resonant llc dc-dc converter. IEEE Trans. Power Electron. **26**(11), 3274–3284 (2011)
23. C.-H. Yang, T.-J. Liang, K.-H. Chen, J.-S. Li, J.-S. Lee, Loss analysis of half-bridge llc resonant converter, in *Future Energy Electronics Conference (IFEEC), 2013 1st International*, (2013), pp. 155–160

Chapter 6
Conclusions and Future Works

This chapter concludes the thesis. Section 6.1 briefly relates the motivation and provides the conclusion on the work done in this thesis. Finally, the direction of future research is discussed in Sect. 6.2.

6.1 Conclusions

Based on specific requirements of the aircraft system, new power electronic converter topologies are presented/proposed in this thesis. For aerospace application, weight is of high concern and therefore, power converters suitable for high switching frequency operation and high power conversion efficiency are proposed, analyzed, designed and implemented.

In Chaps. 2, 3 and 4, matrix (3×1) topology based three phase AC–DC topologies suitable for aircraft system are proposed. Three phase buck rectifiers are preferred over three phase boost rectifier for step down voltage gain because of their higher power density, lower voltage rating of the semiconductor devices, and simpler sensing and closed loop control design. In aircrafts, non-isolated buck rectifiers are ideal choice for front end rectifiers. In Chap. 2, a novel matrix based non-isolated buck rectifier with double step down voltage gain is proposed. The proposed buck rectifier provides half of the output voltage provided by traditional buck rectifier at the same modulation index and thus, promises improved power conversion efficiency and input power quality for large step down output DC voltage. In Sect. 2.5, steady state analysis and design including switch voltage/current stress calculation, accurate voltage/current ripple estimation and effect of modulation index on input current THD are presented in details. Further, in Sect. 2.7, the comparative loss evaluation of the proposed converter with the state of art six switch buck rectifier is carried

© Springer Nature Singapore Pte Ltd. 2018
A. K. Singh, *Analysis and Design of Power Converter Topologies
for Application in Future More Electric Aircraft*, Springer Theses,
https://doi.org/10.1007/978-981-10-8213-9_6

out which shows that the proposed converter provides lower semiconductor loss for 40 kHz switching frequency at 500 W output power. With further increase in the output power, the proposed converter shows significant reduction in the semiconductor loss compared to three switch buck rectifier. A new modified SVM based modulation scheme is proposed and digitally implemented and input current THD < 5% is demonstrated both in simulation and experimental tests. One of the limitations of the proposed topology is higher number of switching devices compared to the conventional buck rectifiers. However, single control of each matrix switches and less number of semiconductor diodes in the proposed topology compensate for higher number of MOSFET switches. Moreover, the switch current always flows through the MOSFET channel similar to synchronous rectification and thus, the losses associated with body diodes are completely eliminated. The proposed nonisolated buck rectifier is benchmarked with respect to passive ATRU and 12.3% improvement in power density and 1.2% improvement in power conversion efficiency are estimated. The proposed topology can be potentially configured as bidirectional buck rectifier by replacing the CDR diodes using controlled semiconductor switches such as MOSFETs.

In Chap. 3, a matrix based isolated AC-DC converter is presented. Being a single stage conversion, the matrix based isolated AC-DC converter is very suitable for high power density. The removal of bulky electrolytic DC link capacitor further improves power density and reliability. It has been estimated in this chapter that the weight of DC link capacitor can be 1.12 kG for 10 kW output power. Moreover, the size and weight of DC link capacitor are independent of switching frequency. In this chapter, the non-isolated matrix converter proposed in Chap. 2 is extended to provide isolated power conversion. However, one of the main challenges in isolated power conversion arises because of the unavoidable leakage inductance of the high frequency transformer. A novel soft-switched active snubber circuit with an additional high frequency series capacitor forming a resonant tank is proposed which overcomes the issue of high frequency current commutation, duty cycle loss and voltage spikes. A reduction of 65% in duty cycle loss is demonstrated through hardware experiments by using the proposed high frequency current commutation method. Further, the presented SVM based modulation scheme is digitally implemented and discussed in details in Sect. 3.7. The proposed scheme is implemented with DSP and FPGA and it provides high frequency and high resolution switching signal for the matrix switches. The performance of the proposed converter topology is demonstrated through both simulation and experimental tests. Further, the proposed converter is compared to existing isolated three phase AC–DC converter and benefits of the proposed converter is discussed. The proposed isolated AC–DC converter is benchmarked with passive TRU and 18% improvement in power density and 3.5% improvement in power conversion efficiency are estimated. In future, new semiconductor devices such as SiC and GaN would be implemented to further improve the performance of the proposed topology.

There is recent trend of increasing input voltage from 115 to 230 V AC for reduced electrical weight in aircrafts. A new matrix based buck-boost converter is proposed in Chap. 5 to provide voltage gain (230–270 V DC) for which none of the traditional

AC–DC rectifiers- buck, boost, buck-boost and boost-buck are suitable. The suitability and limitations of the proposed matrix based buck-boost topology is discussed by comparing it with three switch buck rectifier based buck-boost topology. Steady state analysis and design of the converter is carried out. The design equations are subsequently verified through simulations. A scale down prototype is built to demonstrate the performance of the converter.

In Chap. 5, an LLC resonant converter with SQR circuit is presented. The converter is operated at high switching frequency (230 kHz at full load) with ZVS in MOSFETs and ZCS in SQR diodes contributing to 96% efficiency for 2 kV, 200 W output. The use of SQR circuit reduces the turns ratio of the high frequency transformer by 75% contributing to overall smaller volume and weight. Moreover, the LLC converter is operated below the resonant frequency to get additional voltage boost from the resonant tank. One challenge of operating the converter below resonant frequency is discontinuous tank current for which usual analysis and design based on FHA fail to provide accurate results. In Sect. 5.5.2, the FHA analysis of the LLC converter with SQR is carried out. Further, a new differential based method is proposed for accurate analysis and design of the presented converter and the efficacy and accuracy of this method is validated through simulation and experimental tests which are discussed in Sect. 5.7.4.

Further, the findings of this research work is presented in various peer reviewed journals and conferences. The list of publications is provided.

6.2 Future Work

This research work is mainly focused on proposing new power converter topologies suitable for aircraft system with improved switching modulation schemes. Subsequently, the comprehensive analysis and design are provided for each topology. Further, the proposed topologies are validated through simulation and hardware prototype in a laboratory environment which promise better performance than the conventional power converters.

In future research work, the new semiconductor devices such as SiC/GaN can be implemented and performance improvement needs to be evaluated. The proposed topologies are very much suitable for SiC switches which exhibits excellent Figure Of Merit (FOM) but has large diode forward voltage drop. However, to fully utilize the benefits of SiC switches, the PCB design with low parasitics is of paramount importance. Further studies are required to design PCB layout with smaller current loop and low parasitics to achieve higher switching frequency operation.

Another area of research work can be the parameter optimization. The increased switching frequency reduces passive size and volume but it also increases switching loss resulting in large heat-sinks. These two contradictory requirements must be optimized with consideration of PCB parasitics to have optimum converter design. In the proposed modulation scheme, the switching sequence of matrix switches can have different combinations. In future, different switching sequence for the proposed modulation scheme can be studied and consequently, optimum switching sequence can be investigated for further improvement in power conversion efficiency and input power quality.

The closed loop control of the proposed converters for regulated output DC voltage can be simply implemented using traditional two loop PI control as the output configurations of the converters are same as the traditional converters. However, new control methods such as Model Predictive Control (MPC) can be investigated for further improvement in converter's dynamics. The proposed topologies in Chaps. 2 and 3 are very much suited for bidirectional power flow. The modulation scheme presented in this thesis can be extended for facilitating bidirectional power flow for both isolated and non-isolated AC–DC power conversion. The proposed isolated topology in Chap. 3 can be further extended for energy storage application where bidirectional power flow with galvanic isolation is required between utility grid (AC) and battery storage (DC).

Printed by Printforce, the Netherlands